Praise for Huckleberry Delights

A Collection of Huckleberry Recipes
Cookbook Delights Series Book 6

…"Huckleberries have to be good for you and all these recipes make them better. Have a great excursion picking berries this season and use *Huckleberry Delights* to give your family a treat"…

Bob Howdy, PhD

…"This little berry gets complete and undivided attention in *Huckleberry Delights*, part of the author's Delights series of cookbooks…Hood makes this little berry fun to eat, and she shows an active imagination when it comes to creating recipes"…

Mark Williams

…"Huckleberry connoisseurs, your cookbook has arrived! Karen Jean Matsko Hood has managed to condense more recipes, nutrition, and cultivation information on these pages than any other collection I've seen. ANYTHING and EVERYTHING you've ever wanted to know about huckleberries is in this book. It's a true Northwest gem that's educational and makes entertaining more fun and flavorful!

Huckleberry Delights Cookbook is fresh, savory, and satisfying!"…

Kimberly Carter
Editor

Praise for Huckleberry Delights

A Collection of Huckleberry Recipes

Cookbook Delights Series Book 6

…"Huckleberries are unique to Washington, Idaho, and Montana where they are enjoyed as a delicacy. *Huckleberry Delights Cookbook* brings out the best in huckleberries, using their distinct flavor in well-written and easy-to-read recipes. Karen Jean Matsko Hood has outdone herself in gathering an assortment of information and recipes that make scrumptious and unique dishes worthy of the huckleberry. *Huckleberry Delights Cookbook* is a must-have for any huckleberry lover!"…

> *Mary Scripture-Smith*
> *Graphic Designer*

Huckleberry Delights was featured on "Good Morning Northwest" on September 7, 2008. The live news program is broadcast on KXLY Channel 4, the ABC TV affiliate in Spokane, Washington. Author Karen Jean Matsko Hood presented an array of cookbook recipes which were posted on news4.com. The segment also featured a live cooking demonstration.

Huckleberry Delights

A Collection of Huckleberry Recipes

Cookbook Delights Series Book 6

Karen Jean Matsko Hood

Current and Future Cookbooks
By Karen Jean Matsko Hood

Jicama Delights
Kale Delights
Kiwi Delights
Kohlrabi Delights
Lavender Delights
Leek Delights
Lemon Delights
Lentil Delights
Lettuce Delights
Lime Delights
Lingonberry Delights
Lobster Delights
Loganberry Delights
Macadamia Nut Delights
Mango Delights
Marionberry Delights
Milk Delights
Mint Delights
Miso Delights
Mushroom Delights
Mussel Delights
Nectarine Delights
Oatmeal Delights
Olive Delights
Onion Delights
Orange Delights
Oregon Berry Delights
Oyster Delights
Papaya Delights
Parsley Delights
Parsnip Delights
Pea Delights
Peach Delights
Peanut Delights
Pear Delights
Pecan Delights
Pepper Delights
Persimmon Delights
Pine Nut Delights
Pineapple Delights
Pistachio Delights
Plum Delights
Pomegranate Delights
Pomelo Delights

Popcorn Delights
Poppy Seed Delights
Pork Delights
Potato Delights
Prickly Pear Cactus Delights
Prune Delights
Pumpkin Delights
Quince Delights
Radish Delights
Raisin Delights
Raspberry Delights
Rhubarb Delights
Rice Delights
Rose Delights
Rosemary Delights
Rutabaga Delights
Salmon Delights
Salmonberry Delights
Salsify Delights
Savory Delights
Scallop Delights
Sea Vegetable Delights
Seaweed Delights
Serviceberry Delights
Sesame Delights
Shallot Delights
Shrimp Delights
Soybean Delights
Spinach Delights
Squash Delights
Star Fruit Delights
Strawberry Delights
Sunflower Seed Delights
Sweet Potato Delights
Swiss Chard Delights
Tangerine Delights
Tapioca Delights
Tayberry Delights
Tea Delights
Teaberry Delights
Thimbleberry Delights
Tofu Delights
Tomatillo Delights
Tomato Delights

Trout Delights
Truffle Delights
Tuna Delights
Turkey Delights
Turmeric Delights
Turnip Delights
Vanilla Delights
Walnut Delights
Wasabi Delights
Watermelon Delights
Wheat Delights
Wild Rice Delights
Yam Delights
Yogurt Delights
Zucchini Delights

CITY DELIGHTS
Chicago Delights
Coeur d'Alene Delights
Great Falls Delights
Honolulu Delights
Minneapolis Delights
Phoenix Delights
Portland Delights
Sandpoint Delights
Scottsdale Delights
Seattle Delights
Spokane Delights
St. Cloud Delights

FOSTER CARE
Foster Children Cookbook
  and Activity Book
Foster Children's Favorite
Recipes
Holiday Cookbook for Foster
  Families

GENERAL THEME
  DELIGHTS
Appetizer Delights
Baby Food Delights
Barbeque Delights
Beer-Making Delights
Beverage Delights

Biscotti Delights
Bisque Delights
Blender Delights
Bread Delights
Bread Maker Delights
Breakfast Delights
Brunch Delights
Cake Delights
Campfire Food Delights
Candy Delights
Canned Food Delights
Cast Iron Delights
Cheesecake Delights
Chili Delights
Chowder Delights
Cocktail Delights
College Cooking Delights
Comfort Food Delights
Cookie Delights
Cooking for One Delights
Cooking for Two Delights
Cracker Delights
Crepe Delights
Crockpot Delights
Dairy Delights
Dehydrated Food Delights
Dessert Delights
Dinner Delights
Dutch Oven Delights
Foil Delights
Fondue Delights
Food Processor Delights
Fried Food Delights
Frozen Food Delights
Fruit Delights
Gelatin Delights
Grilled Delights
Hiking Food Delights
Ice Cream Delights
Juice Delights
Kid's Delights
Kosher Diet Delights
Liqueur-Making Delights

Liqueurs and Spirits Delights
Lunch Delights
Marinade Delights
Microwave Delights
Milk Shake and Malt Delights
Panini Delights
Pasta Delights
Pesto Delights
Phyllo Delights
Pickled Food Delights
Picnic Food Delights
Pizza Delights
Preserved Delights
Pudding and Custard Delights
Quiche Delights
Quick Mix Delights
Rainbow Delights
Salad Delights
Salsa Delights
Sandwich Delights
Sea Vegetable Delights
Seafood Delights
Smoothie Delights
Snack Delights
Soup Delights
Supper Delights
Tart Delights
Torte Delights
Tropical Delights
Vegan Delights
Vegetable Delights
Vegetarian Delights
Vinegar Delights
Wildflower Delights
Wine Delights
Winemaking Delights
Wok Delights

GIFTS-IN-A-JAR SERIES
Beverage Gifts-in-a-Jar
Christmas Gifts-in-a-Jar
Cookie Gifts-in-a-Jar
Gifts-in-a-Jar
Gifts-in-a-Jar Catholic

Gifts-in-a-Jar Christian
Holiday Gifts-in-a-Jar
Soup Gifts-in-a-Jar

HEALTH-RELATED DELIGHTS
Achalasia Diet Delights
Adrenal Health Diet Delights
Anti-Acid Reflux Diet Delights
Anti-Cancer Diet Delights
Anti-Inflammation Diet Delights
Anti-Stress Diet Delights
Arthritis Diet Delights
Bone Health Diet Delights
Diabetic Diet Delights
Fibromyalgia Diet Delights
Gluten-Free Diet Delights
Healthy Breath Diet Delights
Healthy Digestion Diet Delights
Healthy Heart Diet Delights
Healthy Skin Diet Delights
Healthy Teeth Diet Delights
High-Fiber Diet Delights
High-Iodine Diet Delights
High-Protein Diet Delights
Immune Health Diet Delights
Kidney Health Diet Delights
Lactose-Free Diet Delights
Liquid Diet Delights
Liver Health Diet Delights
Low-Calorie Diet Delights
Low-Carb Diet Delights
Low-Fat Diet Delights
Low-Sodium Diet Delights
Low-Sugar Diet Delights
Lymphoma Health Support Diet Delights
Multiple Sclerosis Healthy Diet Delights
No Flour No Sugar Diet Delights
Organic Food Delights
pH-Friendly Diet Delights
Pregnancy Diet Delights
Raw Food Diet Delights

Australian Delights
Austrian Delights
Brazilian Delights
Canadian Delights
Chilean Delights
Chinese Delights
Czechoslovakian Delights
English Delights
Ethiopian Delights
Fijian Delights
French Delights
German Delights
Greek Delights
Hungarian Delights
Icelandic Delights
Indian Delights
Irish Delights
Italian Delights
Korean Delights
Mexican Delights
Native American Delights
Polish Delights
Russian Delights
Scottish Delights
Slovenian Delights
Swedish Delights
Thai Delights
The Netherlands Delights
Yugoslavian Delights
Zambian Delights

REGIONAL DELIGHTS
Glacier National Park Delights
Northwest Regional Delights
Oregon Coast Delights
Schweitzer Mountain Delights
Southwest Regional Delights
Tropical Delights
Washington Wine Country
Delights
Wine Delights of Walla Walla
  Wineries
Yellowstone National Park
  Delights

SEASONAL DELIGHTS
Autumn Harvest Delights
Spring Harvest Delights
Summer Harvest Delights
Winter Harvest Delights

SPECIAL EVENTS DELIGHTS
Birthday Delights
Coffee Klatch Delights
Super Bowl Delights
Tea Time Delights

STATE DELIGHTS
Alaska Delights
Arizona Delights
Georgia Delights
Hawaii Delights
Idaho Delights
Illinois Delights
Iowa Delights
Louisiana Delights
Minnesota Delights
Montana Delights
North Dakota Delights
Oregon Delights
South Dakota Delights
Texas Delights
Washington Delights

U.S. TERRITORIES DELIGHTS
Cruzan Delights
U.S. Virgin Island Delights

BILINGUAL DELIGHTS SERIES
Apple Delights, English-
  French Edition
Apple Delights, English-
  Russian Edition
Apple Delights, English-
  Spanish Edition
Huckleberry Delights,
  English-French Edition

Huckleberry Delights,
English-Russian Edition
Huckleberry Delights,
English-Spanish Edition

CATHOLIC DELIGHTS SERIES

Apple Delights Catholic
Coffee Delights Catholic
Easter Delights Catholic
Huckleberry Delights Catholic
Tea Delights Catholic

CATHOLIC BILINGUAL SERIES

Apple Delights Catholic,
English-French Edition
Apple Delights Catholic,
English-Russian Edition
Apple Delights Catholic,
English-Spanish Edition
Huckleberry Delights Catholic,
English-Spanish Edition

CHRISTIAN DELIGHTS SERIES

Apple Delights Christian
Coffee Delights Christian
Easter Delights Christian
Huckleberry Delights Christian
Tea Delights Christian

CHRISTIAN BILINGUAL SERIES

Apple Delights Christian,
English-French Edition
Apple Delights Christian,
English-Russian Edition
Apple Delights Christian,
English-Spanish Edition
Huckleberry Delights
Christian, English-Spanish
Edition

FUNDRAISING COOK BOOKS

Ask about our fundraising
cookbooks to help raise
funds for your organization.

The above books are also available in bilingual versions. Please contact Whispering Pine Press International, Inc., for details.

Please note that some books are future books and are currently in production. Please contact us for availability date. Prices are subject to change without notice.

The above list of books is not all-inclusive. For a complete list please visit our website or contact us at:

Whispering Pine Press International, Inc.
Your Northwest Book Publishing Company
P.O. Box 214
Spokane Valley, WA 99037-0214 USA
Phone: (509) 928-8700 | Fax: (509) 922-9949
Email: sales@whisperingpinepress.com
Publisher Websites: www.WhisperingPinePress.com
www.WhisperingPinePressBookstore.com
Blog: www.WhisperingPinePressBlog.com

Huckleberry Delights

A Collection of Huckleberry Recipes

Cookbook Delights Series Book 6

Karen Jean Matsko Hood

Published by:

Whispering Pine Press International, Inc.

Your Northwest Book Publishing Company
P.O. Box 214
Spokane Valley, WA 99037-0214 USA
Phone: (509) 928-8700 | Fax: (509) 922-9949
Email: sales@whisperingpinepress.com
Websites: www.WhisperingPinePress.com
www.WhisperingPinePressBookstore.com
Blog: www.WhisperingPinePressBlog.com
SAN 253-200X
Printed in the U.S.A.

Published by Whispering Pine Press International, Inc.
P.O. Box 214
Spokane Valley, Washington 99037-0214 USA

For sales outside the United States, please contact the Whispering Pine Press International, Inc., International Sales Department.

Manufactured in the United States of America. This paper is acid-free and 100% chlorine free.

Book and Cover Design by Artistic Design Service
P.O. Box 1782
Spokane Valley, WA 99037-1782 USA
www.ArtisticDesignService.com

Library of Congress Number (LCCN): 2014900835

Hood, Karen Jean Matsko
    Title: Huckleberry Delights Cookbook: A Collection of Huckleberry Recipes: Cookbook Delights Series Book 6

        p.   cm.

ISBN: 978-1-59649-102-1   case bound
ISBN: 978-1-59649-385-8   perfect bound
ISBN: 978-1-59649-776-4   spiral bound
ISBN: 978-1-930948-96-9   comb bound
ISBN: 978-1-59210-017-0   E-PDF
ISBN: 978-1-59210-947-0   E-PUB
ISBN: 978-1-59434-862-4   E-PRC

First Edition: January 2014
1. Cookery *(Huckleberry Delights Cookbook: A Collection of Huckleberry Recipes: Cookbook Delights Series Book 6)*   1. Title

Huckleberry Delights Cookbook

A Collection of Huckleberry Recipes
Cookbook Delights Series Book 6

Gift Inscription

To: _____

From: _____

Date: _____

Special Message: _____

_____

_____

_____

_____

_____

_____

_____

It is always nice to receive a personal note to create a special memory.

www.HuckleberryDelights.com
www.WhisperingPinePress.com
www.WhisperingPinePressBookstore.com

Dedications

To my husband and best friend, Jim.

To our seventeen children: Gabriel, Brianne Kristina and her husband Moulik Vinodkumar Kothari, Marissa Kimberly and her husband Kevin Matthew Franck, Janelle Karina and her husband Paul Joseph Turcotte, Mikayla Karlene, Kyler James, Kelsey Katrina, Corbin Joel, Caleb Jerome, Keisha Kalani Hiwot, Devontay Joshua, Kianna Karielle Selam, Rosy Kiara, Mercedes Katherine, Jasmine Khalia Wengel, Cheyenne Krystal, and Annalise Kaylee Marie.

To our grandchildren and foster grandchildren: Courtney, Lorenzo, and Leah.

To my brother, Stephen, and his wife, Karen.

To my husband's ten siblings: Gary, Colleen, John, Dan, Mary, Ray, Ann, Teresa, Barbara, Agnes, and their families.

In loving memory of my mom, who passed away in 2007; my dad, who passed away in 1976; and my sister, Sandy, who passed away due to multiple sclerosis in 1999.

To Sandy's three sons: Monte, Bradley, and Derek. To Monte's wife, Sarah, and their children: Liam, Alice, Charlie, and Samuel and their foster children. To Bradley's wife, Shawnda, and their children: Anton, Isaac, and Isabel.

To our foster children past and present: Krystal, Sara, Rebecca, Janice, Devontay Joshua, Mercedes Katherine, Zha'Nell, Makia, Onna, Cheyenne Krystal, Onna Marie, Nevaeh, and Zada, our future foster children, and all foster children everywhere.

To the Court Appointed Special Advocate (CASA) Volunteer Program in the judicial system which benefits abused and neglected children.

To the Literacy Campaign dedicated to promoting literacy throughout the world.

Acknowledgements

The author would like to acknowledge all those individuals who helped me during my time in writing this book. Appreciation is extended for all their support and effort they put into this project.

Deep gratitude and profound thanks are owed to my husband, Jim, for giving freely of his time and encouragement during this project. Thanks are also owed to my children Gabriel, Brianne Kristina and her husband Moulik Vinodkumar Kothari, Marissa Kimberly and her husband Kevin Matthew Franck, Janelle Karina and her husband Paul Joseph Turcotte, Mikayla Karlene, Kyler James, Kelsey Katrina, Corbin Joel, Caleb Jerome, Keisha Kalani Hiwot, Devontay Joshua, Kianna Karielle Selam, Rosy Kiara, Mercedes Katherine, Jasmine Khalia Wengel, Cheyenne Krystal, and Annalise Kaylee Marie. All of these persons inspired my writing.

Thanks are due to Beverly Koerperich and Sharron Thompson for their assistance in editing and typing this manuscript for publication. Thanks go to Artistic Design Service for their assistance in formatting and providing a graphic design of this manuscript for publication. This project could not have been completed without them.

Many thanks are due to members of my family, all of whom were extremely supportive during the time it took to complete this project. Their patience and support are greatly appreciated.

Huckleberry Delights Cookbook

Table of Contents

Huckleberry Delights Cookbook

A Collection of Huckleberry Recipes
Cookbook Delights Series Book 6

Introduction

Living in the heart of Washington brings great appreciation for the huckleberry season. Huckleberry bushes are beautiful! Daily, we enjoy the beauty of the huckleberry bushes and the many stories they have inspired. Many of us have wonderful memories of picking huckleberries in the hot July and August sun and running away when a bear came to claim his territory.

Huckleberries are indeed a tasty and delicious food. Huckleberry varieties come in many colors, sizes, and textures. The fruit is great for cooking and is nutritious to eat alone. It is no wonder that huckleberry bush cultivation quickly spread throughout the western United States.

Huckleberries have an interesting history of facts and folklore. Some of this huckleberry folklore is included in this book. As a poet, I found it enjoyable to color this cookbook with poetry so that readers could savor the metaphorical richness of the huckleberry as well as its literal flavor. Also included in this *Huckleberry Delights Cookbook* are some articles on history, cultivation, and botanical information, along with interesting facts about huckleberries.

The *Cookbook Delights Series* would not be complete without *Huckleberry Delights* because huckleberries are a common and popular Pacific Northwestern American fruit. We hope you enjoy reading it as well as trying out all the recipes. This cookbook is designed for easy use and is organized into alphabetical sections: appetizers and dips; beverages; breads and rolls; breakfasts; cakes; candies; cookies; desserts; dressings, sauces, and condiments; jams, jellies, and syrups; main dishes; pies; preserving; salads; side dishes; soups; and wines and spirits.

Do enjoy reading about huckleberries, but most importantly, have fun with those you care about while you are cooking.

Following is a collection of recipes gathered and modified to bring you *Huckleberry Delights Cookbook: A Collection of Huckleberry Recipes, Cookbook Delights Series* by Karen Jean Matsko Hood.

Huckleberry Delights Cookbook

A Collection of Huckleberry Recipes
Cookbook Delights Series Book 6

Huckleberry Botanical Classification

Huckleberry Botanical Classification

Scientific Name	Common Names	Main Features
Vaccinium globulare	blue huckleberry globe huckleberry	flowers broader than long, rounded leaf tips; common to Montana
V. membrana-ceum	mountain huckleberry thin-leaved huckleberry mountain bilberry big whortleberry black huckleberry tall bilberry	flowers longer than broad, "drip tip" leaf ends; common to Idaho and Cascade areas
V. caespitosum	dwarf huckleberry	shorter shrubs, under 18 inches, rounded stems; obovate leaves (egg-shaped leaf shape with a narrow tip)
V. myrtilloides	velvet-leaved huckleberry	distinctive hairiness to leaves and stems; common to Alberta, Canada
V. myrtillus	bilberry dwarf bilberry low bilberry	green stem, small shrub, branching but not broom-like, lower elevations than *scoparium*
V. scoparium	grouse whortleberry	tiny bright red berries, green stem, branching broom- like
V. occidentale	western huckleberry	no serration visible on leaf edges, even with hand lens; known to sub-alpine wet meadows near Idaho/Montana border

Huckleberry Delights Cookbook

A Collection of Huckleberry Recipes
Cookbook Delights Series Book 6

Huckleberry Cultivation and Gardening

Huckleberry Cultivation and Gardening

Huckleberries grow best in the wild. In the Northwest many gardeners have tried to cultivate huckleberries with little success. Till rotted sawdust or bark into the soil a year before planting. This can improve huckleberry survival and growth. Include one pound of 10-10-10 fertilizer for every cubic foot (13 pounds) of sawdust or bark.

Huckleberry plants used to outdoor conditions can be planted any time from early spring through late fall. If the soil in your garden has heavy frost, plant in the spring. Immediately after planting, heavily water. Mulch around each huckleberry bush with about four inches of sawdust or fine bark to keep soil moist and protect the roots. At least some huckleberries may form symbiotic relationships with other plants that are helpful. You can provide these fungi for your plants by mixing a shovelful of soil collected from a native huckleberry site with the backfill from each planting hole. When collecting native soil, scrape off the duff layer and collect the soil from the surface to about eight inches deep. Include pieces of buried, rotted wood from the site, if available.

Field-grown huckleberries respond well to granular, liquid, and slow-released fertilizers, as well as manures. Do not use weed and feed fertilizers. Follow label directions for liquid and slow-release fertilizers.

Huckleberries grow slowly. Other than removing dead or damaged branches, pruning is not needed nor recommended in young plantings. It is not yet known how older, cultivated huckleberries respond to pruning. Dense plants, however, may benefit from occasional light pruning that opens the bushes up and improves light penetration and air movement. Although eastern lowbush blueberry fields are burned every two years and fires are common in the forests where huckleberries grow wild, burning is not recommended for western huckleberries. Burned huckleberry fields can take 10 to 15 years to return to full productivity.

Huckleberry Delights Cookbook

A Collection of Huckleberry Recipes

Cookbook Delights Series Book 6

Huckleberry Facts

Huckleberry Facts

Huckleberries have not yet been domesticated and are generally hand-picked in their native habitat. They have a sparse root system which makes transplanting very difficult, and they do not absorb nutrients well. Very close attention has to be given to nutrients and irrigation. They do not propagate well from cuttings, so they need to be started from seed. When started from seed, they may not produce a plant with the parent's characteristics, and they may take 10 years to produce. The huckleberry plant is covered by snow in the mountains it calls home, so at lower elevations there is the possibility of winter kill. The soil type can also be a problem, since they need an acidic, well-drained soil.

Huckleberries begin to ripen in August each year in the Pacific Northwest and have a short season. The huckleberry does not keep long, nor does it ship well. Therefore, it is made into jams and jellies for a saleable product that keeps for months and can be shipped anywhere.

The market price for huckleberries is dependent on the weather. If there is plenty of rainfall during the spring and early summer and enough hot sunny days afterward, the berries will be large and plentiful. Hence, the market price drops due to the abundance of berries available. When the weather conditions are not favorable, the market price for the delicacy inflates.

The Native Americans of the Pacific Northwest harvested these berries. Some tribes dried the whole berries over a smoldering fire, while others cooked and mashed them before drying them in the sun or by wrapping the "jam" in skunk cabbage leaves and drying it over a slow fire. These fruit cakes were then rolled up and stored for later use.

Wild berries, including the huckleberry, are used as a ceremonial food at the First Salmon Celebration each spring. Huckleberries were also used by the Native Americans as a dye. The leaves of the plant were made into tea or smoked, branches made brooms, and twigs fastened skunk cabbage leaves into their berry baskets.

Huckleberry Delights Cookbook

A Collection of Huckleberry Recipes
Cookbook Delights Series Book 6

Huckleberry Folklore

Huckleberry Folklore

Huckleberries have been designated as the official berry of the State of Idaho. There are many family stories that include accounts of facing bears on the other side of the huckleberry bush, and accounts of children's early memories of huckleberry hunts with their family. Some families make the trek into the woods and mountains for the sheer fun and satisfaction of boasting: "I picked a gallon in two hours! They are as big as your thumb, and the bushes are just loaded!" However, if you ask that person the exact location of the huckleberry patch, you are likely to receive a rather icy stare, with inadequate directions. The jealous guarding of the exact location of the "wonderful" patch of huckleberries is a family tradition in the Pacific Northwest. Visitors to the Priest Lake area of Idaho find the berries to be especially tasty and large. In that area, bears are especially dependent on the berries as a carbohydrate source to sustain them during hibernation. Accordingly, if you go to Priest Lake to hunt for berries, always be prepared to meet Madam Black Bear or Sir Grizzly when you are in huckleberry country. It does not occur often, but when it does, the wise picker always moves on to another location and allows the bear to feed without being disturbed.

A disease has been associated with the picking of these berries. The disease has no cure and its symptoms include talking to little purple berries and having purple fingers and tongues. The disease has been called by the not-so-scientific name, "huckleberry mania."

Huckleberry was used in different phrases in the late 19th century. When referring to a person, huckleberry meant that person was the right person for a job. It could also mean the person was just very nice or even a sweetheart. Because of the small size of the berry, huckleberry was also used to mean a small amount of something or refer to a person or thing that was insignificant. Huckleberry was used is phrases like "to bet a huckleberry to a persimmon." This meant that it was a very small bet. "A huckleberry above a persimmon" meant a very small amount of something. But, it was also used to mean something unique in the phrase "the only huckleberry on the bush."

Huckleberry Delights Cookbook

A Collection of Huckleberry Recipes
Cookbook Delights Series Book 6

Huckleberry History

Huckleberry History

The arrival of spring in the Pacific and Inland Northwest brings an abundance of sweet berries that can be found ripening in the wilderness in late summer. The Pacific Northwest is rich in plentiful and assorted groups of shrubs that provide a succulent harvest for berry pickers. The term "huckleberry" is used by most people to identify wild blueberries. Common to areas along the Pacific Coast and inland coniferous forests from Northern California to Alaska, huckleberries have a long history of use by the native people of this area as an important resource for food, drink, and rich-colored dyes. Although a number of species of huckleberries are indigenous to the Pacific Northwest, in the lowland area of Puget Sound, there are two types of huckleberries, the red and evergreen huckleberry, which have remarkably different appearances and produce distinctive fruits.

Both types are delightful additions to home gardens, and are especially appreciated by those wanting to attract more butterflies and birds to their property. Their year-round color makes them especially pleasing in winter and early spring, when many other garden plants may be leafless. While both varieties of huckleberries grow best in full to partial shade, the evergreen huckleberry will withstand the stress of full sun. The red huckleberry is an especially good choice in dry, shady, woodland gardens or mixed with rhododendrons.

In addition to serving as a ready source of food, the huckleberries provided other benefits to native peoples. The red huckleberry's bright fruits were prized as fish bait, its juice was used as a mouthwash, and its leaves and bark were prepared into a medicine for sore throats and inflamed gums. Evergreen huckleberries were, and still are, highly prized for their flavor and the persistence of the fruits into December. In fact, some insist the berries are better after the first frost.

Huckleberry Delights Cookbook

A Collection of Huckleberry Recipes
Cookbook Delights Series Book 6

Huckleberry Nutrition and Health

Huckleberry Nutrition and Health

As the interest in nutrition increases and people are becoming more and more conscious about what they eat, the interest in fruits and berries and their contents also grows. Although we still do not know much about the nutritional content of huckleberries, it is a fruit and, therefore, holds many of the same characteristics that are true for other fruits and berries.

Since the huckleberry is so closely related to the blueberry, it most likely would have very similar nutritional value. Therefore, it would be a very good source of vitamin C and manganese, as well as a good source of vitamin E.

Blueberries are the highest rated among 60 different fruits and vegetables for their antioxidant capabilities. A study also found that blueberry wine delivered 38 percent more antioxidant anthocyanins than red wine, which is touted for its cardioprotective capability.

Studies of the bilberry (a cousin of the blueberry) have shown that its extract improves nighttime visual acuity. And, fruit may be even better for your eyesight than carrots!

There are many good reasons to eat huckleberries and other fresh fruits. First of all, most fruits and berries consist of mostly water, and they are 100% bad cholesterol-free. Eating food that is rich in water content makes it easier for your body to digest, so that your body can use its energy for other purposes, such as thinking. Fruit and berry juices take only about 15 minutes to digest, and raw fruits or berries take about 30 minutes to digest. A steak can take up to 8 to 10 hours to digest, especially when eaten in combination with potatoes. In comparison to fruits and berries (30 minutes), a lot of energy is lost to the digestion process of proteins, and this means that you cannot use that energy for other things like thinking or disposing of wasteful toxins in the body. Fruits and berries stimulate our brains to work clearer and smoother. In fact, it has been said that students eating fruit before an exam found that their test results were higher than when they did not.

Fruits and berries contain healthy fibers, and a diet with plenty of fiber can be helpful in the fight against high blood pressure and colon disease.

Huckleberry Delights Cookbook

A Collection of Huckleberry Recipes

Cookbook Delights Series Book 6

Poetry

A Collection of Poetry with Huckleberry Themes

Table of Contents

Huckleberry Cluster

Shrubs of similar roots;
similar yet different
sway in the wind
now still.
Exclamations of leaves gather
in one spot.

There, in the branches of cover,
shout small spheres of purple
over a few leaves,
now crimson.

To the north,
stand grizzlies.
To the south,
walk humans.
Both in search of purple treasure.
Rocks grumble, branches snap.
Can nature harmonize
in the sunset?
    For one moment . . .
    nothing changes.

Karen Jean Matsko Hood ©2014
Published in *Huckleberry Delights Cookbook*, 2014
By Whispering Pine Press International, Inc., 2014

Huckleberry Memories

Lush, lavender ice cream drips purple pieces,
huckleberry textures, piles high and
perches upon homemade cones.
Children grab while grandparents wait.
Huckleberry candy twists and
spirals on purple mounds that punctuate the ice crystals.
Delicious purple memories, the royal
display of colors, fragrant in
our childhood minds.

Karen Jean Matsko Hood ©2014
Published in *Huckleberry Delights Cookbook*, 2014
By Whispering Pine Press International, Inc., 2014

Huckleberry Petals

White petals unfold to reveal
blue-purple berries in bunches.
Visitors pick clusters on the mountain,
steep inclines in the sky.
Some look over their shoulder
and invade the space of the secret forests
to search for bountiful, juicy huckleberries.

From heavy branches, they pick with wrinkled hands
and seek out each succulent berry.
One by one they fill their buckets
to the top. The fruit line
never seems to reach the rim as
they dip and taste each savory drop
faster than tired hands can fill.
Sweet purple treats,
all from white petals.

Karen Jean Matsko Hood ©2014
Published in *Huckleberry Delights Cookbook*, 2014
By Whispering Pine Press International, Inc., 2014

Huckleberry Blossoms

Huckleberries blossom with
        Telegrams of love that flourish,
      Elegant calligraphy expressed by each petal
       Reminds me of the man I love.

Each spring my beloved gets so excited
        To watch the huckleberries ripen
      Like suspended white ornaments
        On emerging hills of green.

White blossoms replace the withering snow and
        Trade ice for the warmth of spring.
      Lacy blossoms peek out from twigs,
        Rainy cold still looms.

Daffodils smile their friendly greeting,
        And hummingbirds return.
      Down the road the rhubarb sprouts.
        Hyacinths perfume their song.

Muddy fields wait to dry
        As gardeners select their seed.
      Huckleberries blossom on the hillsides,
        Upland fields adorned with splotches
of white;

        Spring communiqué to my love.

Karen Jean Matsko Hood ©2014
Published in *Huckleberry Delights Cookbook*, 2014
By Whispering Pine Press International, Inc., 2014

Huckleberries

Small, dark, violet berries cluster on the bush.
Fragrant, delectable berries ripen to burst.
One by one, slowly, over the summer season
huckleberry leaves form a tapestry
on the forest floor. The palette
blends its colors over mountain slopes which
rise to peaks that touch white clouds.
Pickers travel many miles to search.
Purple bounty tucked under forest canopies
hides in rugged terrain. Each pair
of hands works to fill their bucket
of pure royal stash valued by
those that crave its mountain
berry flavor. Regional pride takes
over as each keeps his or her wild patches
secret. No one wants to whisper
where they find their sweet, savory,
violet gold, distilled juices packaged
perfectly on the bush. Jewels of
lividity hide under Northwest sun.

Karen Jean Matsko Hood ©2014
Published in *Huckleberry Delights Cookbook*, 2014
By Whispering Pine Press International, Inc., 2014

Driving Through Missoula

In a light blue car I sat
    on cracked tan vinyl seats
to sleep through sunset hours and
    drift along the highway,
as peaks come to life.
    Mountains told their story
slow and precise with each passing mile.
    Ponderosa pines jump
slowly up, then down.
Take their turn to talk.
    Quiet yellow buttercups
were wise indeed. They send
huckleberry plants that run
    to hide and keep their bounty treasure
    a secret.

Karen Jean Matsko Hood ©2014
Published in *Huckleberry Delights Cookbook*, 2014
By Whispering Pine Press International, Inc., 2014

Grizzly's Huckleberries

Can you imagine the huge grizzly
with long claws and bulky bear feet
that picks those tiny huckleberries?

How do they pick? How do they manage?
Think of their huge stomachs,
ravenous appetites.

How can they fill the vacancies
with tiny scarce berries?
Treasure jewels of the Northwest?

Candy treat of the grizzly.
Do they sort the berries from leaves
or do they simply make a salad?

Yes, I'd like to see those
hungry grizzlies,
the humps on their backs.

They pick berries to satisfy their sweet tooth.
I'd like to be off in a distance
not too very far away.

Karen Jean Matsko Hood ©2014
Published in *Huckleberry Delights Cookbook*, 2014
By Whispering Pine Press International, Inc., 2014

Buffalo Hair

Patches of hair float from the old buffalo,
once brown, not bleached blonde
under the sun. Silky soft in
silence. Strands hang from the stems
of the bush that dies as
the fox runs past. Fat prairie hens
chuckle as pairs of grouse
mate under the huckleberry bush,
quiet in the forest. Down around
the bend in the road was the tall
buffalo's shrunken bones under
the fur of his cape, pushed off the
cliff, young and old do not survive.

Karen Jean Matsko Hood ©2014
Published in *Huckleberry Delights Cookbook*, 2014
By Whispering Pine Press International, Inc., 2014

Logan Creek

When water weeps from the rock wall
    and snowdrifts melt to become mounds,
Grizzly bears eat to fill their bellies
    and huckleberries fade into russet glow.
It is the time when night hawks soar.

Tundra hardens with each coming frost
    as fat swells on belly after belly.
Birds sing but some forget the notes
    or is it that the ears no longer remember.
Waxy glacial lily petals long gone.

Karen Jean Matsko Hood ©*2014*
Published in *Huckleberry Delights Cookbook*, 2014
By Whispering Pine Press International, Inc., 2014

Huckleberry Delights Cookbook

A Collection of Huckleberry Recipes

Huckleberry Types

Huckleberry Types

Huckleberries and their relatives are placed in the *Ericaceae,* or Heath, family (one of the most widespread and interesting plant families in the world). Commercial blueberries and cranberries also belong to the same "family" as huckleberries. Members of the *Ericaceae* can be shrubs, prostrate herbs (such as heather), tropical vines, and even trees (such as our native Pacific Madrone). Many of the tropical relatives in this family grow epiphytically as shrubs, and their flowers are important food sources for arboreal birds.

Both types of our local huckleberries have drooping, bell-shaped flowers, usually white to pink. Both types range in size from 3 to 12 feet tall—tallest in deep shade and dwarfed in full sun.

The red huckleberry is deciduous, with small (about ½ inch), delicate leaves, while evergreen huckleberries have thick, shiny leaves that densely cover the branches, as its name implies. Even after shedding its leaves, the red huckleberry adds brilliant color to the forest with its jade-green, strongly angled stems covered with small, bright pink buds. In March and April the new leaves of the evergreen huckleberry emerge in a burst of brilliant bronze color at the tips of the branches.

Red huckleberry bushes produce dazzling ruby berries, about ¼ inch in diameter. These berries have a wonderful flavor, though some may say that they are entirely too tart to eat straight from the bush. The evergreen huckleberry's fruits are far sweeter, shiny black, and buckshot-sized. Look for red huckleberries in early to midsummer and the evergreen's berries in late summer and early autumn.

Our local huckleberries are common in the Western Lowlands throughout the Pacific Northwest from British Columbia through Oregon and even into parts of central California.

Both shrubs grow in the moist to slightly dry, shady, coniferous areas, often at forest edges. The evergreen huckleberry is also found in the salt-spray zones and is abundant in the Olympic and

Kitsap Peninsulas. Red huckleberries are renowned for their habit of growing from old stumps and rotting logs.

Approximately 200 of the shrubs which live in the cold mountainous areas of the world have been classified under the genus *Vaccinium*. About 20 species of these shrubs are native to Canada and the northern United States, and they can be found most often at elevations greater than 2,500 feet on the coast (4,000 feet in the interior). You should see the way that huckleberries paint an entire mountain red during the fall—what a breathtaking vision!

The shrubs, which are quite bushy, range in height from 5 feet at lower elevations to 1 foot near timberline. They are crooked plants with smooth limbs and slightly angled twigs. As for the leaves, they are pointed at both ends. They are also finely toothed and noticeably paler on the underside. The leaves range in length from ½ inch to 2 inches.

The berry that is most often scouted out is the black mountain huckleberry, *Vaccinium membranaceun,* which many pickers consider the finest and tastiest of all.

Vaccinium ovalifolium (also common to the area) is the tall, blue huckleberry which grows on bushes. Often these bushes reach 4 feet high. However, the blues are not nearly as sweet as the blacks, and their seeds are much harder.

What is the difference between huckleberries and blueberries? As a member of the genus *Vaccinium*, the huckleberry is related to the wild blueberry of Michigan, Maine, and other points east. In scientific terms it is equally close to the cultivated blueberry found in cans at the grocery store, muffins at the corner bakery, and prefabricated snack pies all over the world. But here the kinship ends and botanical family ties come undone, as far as most berry lovers are concerned, because huckleberries and blueberries simply do not taste the same.

A true huckleberry advocate will usually dismiss blueberries, wild or cultivated, as "blahberries." Others say the two are as much alike as sweet cherries and pie cherries, or a fresh Granny Smith apple compared to last year's Golden

Delicious. But blueberry fanciers are equally loyal, and will reject western varieties as sour imposters for the "true blue." So while a huckleberry may have more in common with a blueberry than it does with, say, a rocking chair, any educated tongue is going to know which you are putting on the toast.

Taste is not the only difference between blueberries and huckleberries. Most species of blueberries grow in bunches, and huckleberries singly, along the stem. Huckleberry bushes tend to have shallow roots with rhizomes, by which they reproduce. Blueberry roots go deeper.

However, there are species of huckleberry that bear fruit in bunches, and roots of the blueberry may grow deep or shallow depending on the soil and other local factors. Such blurred distinctions are quite common among the *Vaccinium*, causing endless annoyance for the scientist wanting to make firm identification of a plant.

The blueberry-huckleberry controversy may be best understood as a family quarrel. Both belong to *Ericacae*, the heath family. This designation tells a botanist that the plants have more in common with one another than they do with mint, geranium, cactus, palm, rose, or other family groups.

The heath or heathers are woody plants with leathery leaves and bilaterally symmetrical flowers; all have twice as many stamens as petals, and upright anthers.

Within the large *Ericacease* family, huckleberries and blueberries are related most closely to one another, which may account for their sibling rivalry. Even *Gaylussacia* become a stepsister in their small sub-family. Consider that the next of kin to both the huckleberry and blueberry is the cranberry. Botanically speaking, cranberries are as much like the high mountain huckleberry as the store pie blueberry. The blueberry is to the huckleberry as the cranberry is to the blueberry.

Huckleberry Delights Cookbook

A Collection of Huckleberry Recipes
Cookbook Delights Series Book 6

RECIPES

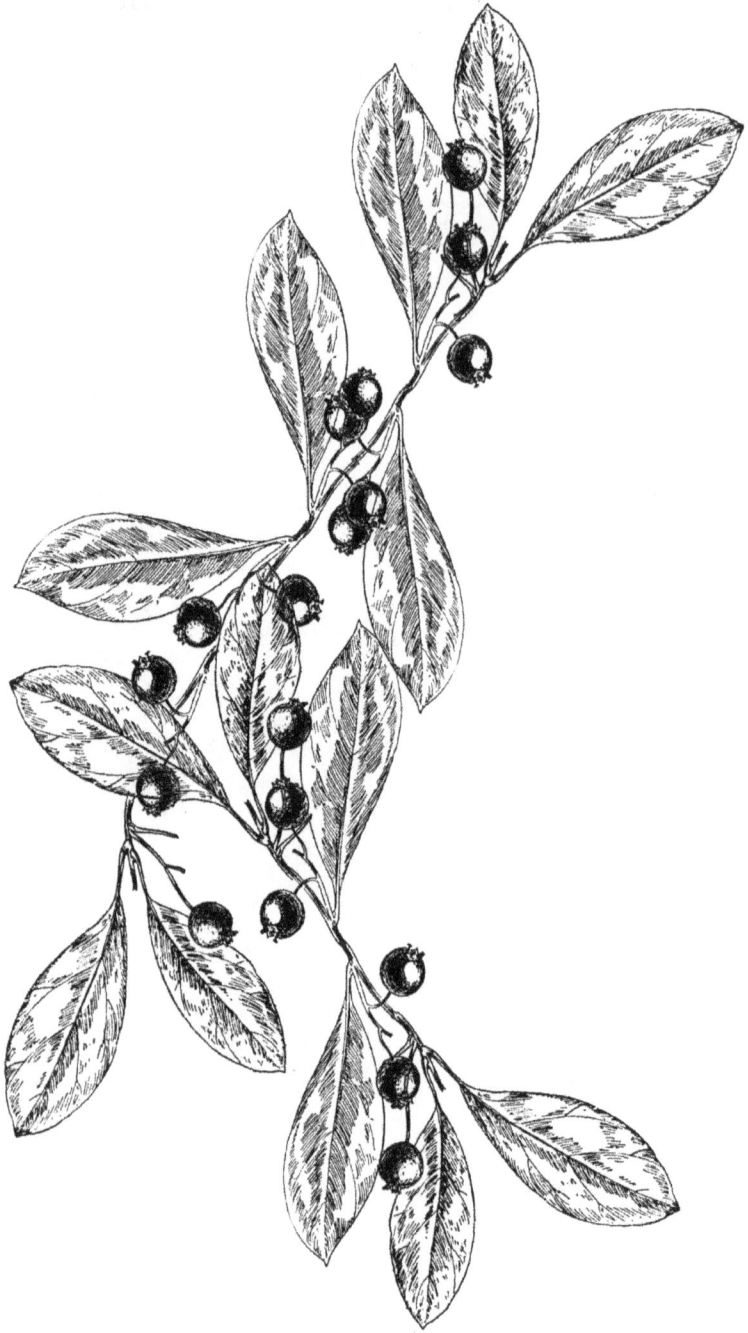

Huckleberry Delights Cookbook

A Collection of Huckleberry Recipes
Cookbook Delights Series Book 6

Appetizers and Dips

Table of Contents

Page

Antipasto with Prosciutto, Huckleberries, and Balsamic Onions

This is a delicious summertime appetizer, perfect for using your freshly picked huckleberries. Do not forget to start your preparations the day before you need the dish.

Ingredients:

1	Tbs. olive oil
2	lg. red onions, peeled, cut in half, thinly sliced
3	Tbs. balsamic vinegar, divided
½	lb. prosciutto ham, thinly sliced
1½	lb. huckleberries, washed
⅓	lb. Parmesan cheese or Asiago cheese, thinly sliced

Directions:

1. The day before serving, heat olive oil over medium-high heat in large skillet.
2. Add onions, and quickly sauté just until they begin to soften, about 4 minutes.
3. Stir in 2 tablespoons balsamic vinegar, stirring to coat; transfer mixture to bowl; cover and refrigerate.
4. Two hours before serving remove onions from refrigeration, and stir in remaining vinegar.
5. Roll each slice of prosciutto loosely, and arrange at one end of large platter; pile huckleberries in center and onions at other end.
6. Tuck slices of cheese around platter, and serve.

Yields: 4 to 6 servings.

Did You Know? . . . .

Did you know that Washington, Idaho, and Montana are the top three huckleberry-producing states in the nation?

Huckleberry Chicken Salsa Torte

This is an impressive appetizer, and finding the huckleberry salsa at your specialty grocer or making it yourself will be well worth the effort.

Ingredients:

- 1   Tbs. olive oil
- 1   sm. onion, cut into strips
- 2   cloves garlic, minced
- 3   c. zucchini, grated
- ¾ lb. canned white meat chicken, drained, shredded
- 3   c. Monterey Jack cheese, shredded
- 3   flour tortillas
- 16 oz. huckleberry salsa (or recipe page 174)
-    sour cream (optional)

Directions:

1. In large skillet heat oil, then add onion and garlic; sauté for 5 minutes.
2. Add zucchini, and sauté for another 5 minutes, stirring occasionally.
3. Remove skillet from heat and drain well; stir in drained, shredded chicken; set aside.
4. Preheat oven to 400 degrees F.
5. Spray 10-inch pie plate with cooking spray.
6. Spread half of chicken mixture into it, then sprinkle with half the cheese.
7. Place 1 tortilla on top of cheese layer, then spread on half the salsa and add 1 more tortilla.
8. Spread on remaining half of salsa then remaining half of chicken mixture over tortilla.
9. Top with 1 more tortilla, and sprinkle with remaining half of cheese.
10. Cover with foil and bake for 40 minutes.
11. Remove cover, and bake for an additional 15 minutes.
12. Remove from oven and let cool for 10 minutes.
13. Cut into 6 or 8 wedges, and serve with small dollops of sour cream.

Yields: 6 to 8 servings.

Chinese Huckleberry Barbequed Pork

This version has additional spice and flavoring, but it is just as delicious, if not more so, than the traditional barbeque!

Ingredients:

¼ c. soy sauce
1   Tbs. red wine
1   Tbs. brown sugar, firmly packed
2   Tbs. huckleberry preserves (recipe page 197)
1   tsp. red food coloring (optional)
1   tsp. blue food coloring (optional)
½   tsp. ground cinnamon
1   green onion, halved
1   clove garlic, minced
2   pork tenderloins (12 oz. each)
    additional huckleberry preserves for dipping
    sesame seeds for dipping

Directions:

1. Combine soy sauce, wine, brown sugar, huckleberry preserves, food colorings, cinnamon, onion, and garlic in large bowl.
2. Trim all fat from meat, and add meat to mixture in large bowl, turning to coat completely.
3. Cover and refrigerate 1 hour or overnight, turning meat occasionally; reserve marinade.
4. Preheat oven to 350 degrees F.
5. Place meat on wire rack in baking pan; bake for 45 minutes or until no longer pink in center, basting frequently with marinade.
6. Chill cooked meat in refrigerator until room temperature or slightly chilled.
7. Slice diagonally, and serve with dish of melted huckleberry preserves and a dish of sesame seeds alongside for dipping.

Chocolate Huckleberry Cheese Ball

Try this unique appetizer for a change of pace from the traditional cheese ball. This sweet cheese ball is great to serve with hot chocolate or flavored coffee on a chilly day. This is also good spread on toasted bagels.

Ingredients:

- 6 oz. semisweet chocolate chips
- 1 c. pecans
- ½ c. dried huckleberries (recipe page 238)
- 8 oz. cream cheese, softened
  chocolate-flavored or vanilla wafers, chocolate or graham crackers, or your favorite sweetened cookie or cracker

Directions:

1. Process chocolate chips and pecans in blender or food processor until finely ground.
2. Place mixture in airtight plastic bag or container until ready to use. (There will be enough mixture to make several cheese balls.)
3. When ready to make cheese ball, place cream cheese in bowl.
4. Remove ½ cup of ground chocolate/nut mixture from its bag, and mix into cream cheese along with dried huckleberries.
5. Shape into ball and wrap in plastic wrap.
6. Refrigerate for 2 or 3 hours or until firm.
7. Serve with chocolate-flavored or vanilla wafers, chocolate or graham crackers, or your favorite sweetened cookie or cracker.

Did You Know? . . . .

Did you know that "bilberry" is another name for the huckleberry?

Festive Nut Bowl

This is an easy and quick-to-fix snack. You can add your favorite granola cereal or exchange some nuts with dried fruits for variations.

Ingredients:

- ½ c. macadamia nuts
- ½ c. salted cashews
- ½ c. shelled pistachio nuts
- ½ c. dried huckleberries (recipe page 238)
- ½ c. dried cranberries
- ½ c. dried cherries

Directions:

1. In large bowl mix together macadamia nuts, cashews, and pistachios.
2. Add dried huckleberries, cranberries, and cherries; toss well to mix.
3. Store in sealed bags until needed or for up to 1 month.
4. To serve, place in attractive serving bowls with small serving spoons.

Yields: 3 cups.

Chocolate Huckleberry Zinfandel Pâté

This is truly a decadent appetizer, and it will definitely whet the appetite. Try it when serving a pork or poultry main course.

Ingredients:

- 1 lb. bittersweet chocolate, chopped
- ¾ c. zinfandel or other red wine
- ¼ c. heavy whipping cream
- 2 pt. fresh huckleberries
- ½ c. sugar

Directions:

1. Combine chocolate, wine, and cream in top of double boiler, and melt over simmering water over low heat, stirring constantly until mixture is smooth. (Do not allow chocolate to get too warm.)
2. Remove from heat and whisk well.
3. Pour into 8 x 4-inch loaf pan lined with wax paper; refrigerate overnight.
4. Before serving, unmold and slice with knife dipped in hot water.
5. Purée berries and sugar in blender until smooth.
6. Strain purée through cheesecloth or fine-mesh sieve if necessary to remove any pieces or chunks.
7. To serve, make pool of sauce on each of 8 serving plates.
8. Slice pâté, and place a slice on top of sauce on each plate.

Yields: 8 servings.

Frosted Huckleberries

These little treats are yummy, and they look nice on your buffet table or served alongside sliced cheese and crackers.

Ingredients:

2 lb. extra-large huckleberries
3 oz. blueberry or grape-flavored gelatin mix

Directions:

1. Rinse huckleberries in colander and drain off excess water; blot lightly to almost dry.
2. Pour gelatin mix into bowl; add one handful of berries at a time, and roll around until coated.
3. Transfer to attractive serving dish, and refrigerate for 1 hour to allow gelatin to set before serving.

Huckleberry Wings

My sons love chicken wings, and this is an easy variation of the classic chicken wings recipes.

Ingredients:

1½ c. huckleberry preserves (recipe page 197)
⅓ c. balsamic vinegar
3 Tbs. soy sauce
1½ tsp. crushed red pepper
5 lb. chicken wings

Directions:

1. Preheat oven to 375 degrees F.
2. Line two baking sheets with aluminum foil.
3. In small saucepan combine preserves, vinegar, soy sauce, and red pepper; stir over medium heat until well blended, then remove from heat and cool.
4. In large bowl toss wings with half the preserves mixture, then place on baking sheets; bake for 20 minutes.
5. Turn wings and brush with remaining preserves mixture.
6. Bake for 8 to 10 minutes more or until no pink remains in chicken and sauce glazes wings; serve immediately.

Yields: 6 servings.

Sweet Kielbasa for a Crowd

This is a delicious appetizer for a crowd that you can make earlier in the day in your crockpot. Once it has cooked on high, you can just turn down the heat and let it simmer the rest of the day until needed. Be sure to have a dish of toothpicks handy for eating utensils.

Ingredients:

4 pkg. fully cooked kielbasa sausages (16 oz. each)
1 c. huckleberry preserves (recipe page 197)
2 c. jellied cranberry sauce
1 can crushed pineapple in heavy syrup (20 oz.)

Directions:

1. Preheat oven to 350 degrees F.
2. Cut kielbasa rings into ¼-inch slices.
3. Place sliced kielbasa into 13 x 9 x 2-inch baking dish.
4. In medium bowl stir together huckleberry preserves, cranberry sauce, and crushed pineapple in syrup.
5. Pour over kielbasa and stir to coat.
6. Bake for about 1 hour or until sauce thickens.
7. Note: May also be cooked in crockpot following same directions but allowing additional time for sauce to thicken. Be sure to turn down to low heat once sauce thickens.
8. Serve with crackers or sliced baguette rounds.

Warmed Brie with Huckleberry Chutney

Our family loves brie cheese, and this is one of our favorite ways to serve it.

Ingredients:

1	c. huckleberries, fresh or frozen
2	Tbs. chopped onion
1½	tsp. gingerroot, grated
¼	c. brown sugar, firmly packed
2	Tbs. cider vinegar
1½	tsp. cornstarch
⅛	tsp. salt
1	cinnamon stick
1	brie round, 8 or 12 in.

Directions:

1. In saucepan combine all ingredients except brie; mix well.
2. Bring to a boil over medium heat, stirring frequently, and boil 1 minute.
3. Remove cinnamon stick; cover and refrigerate at least 30 minutes.
4. Place brie on microwave-safe serving dish; microwave on high for 2 to 3 minutes or until warm; remove from oven.
5. Top cheese with cold chutney.
6. Serve with crackers or a sliced baguette.

Warm and Cheesy Fruit Dip

This blend of cheese and fruit creates a wonderful flavor. Although it can be served with crackers, it is especially wonderful served with sliced apples and pears.

Ingredients:

      16 oz. soft-style cream cheese with pineapple, softened
      ¾   lb. Swiss cheese, shredded
      2   c. dried huckleberries (recipe page 238)
      2   Tbs. orange juice
      ¼   c. apple juice

Directions:

   1. Preheat oven to 375 degrees F.
   2. In medium bowl blend cream cheese, Swiss cheese, huckleberries, orange juice, and apple juice.
   3. Scoop into 9-inch pie pan.
   4. Bake until bubbly and lightly browned, about 15 minutes.
   5. Remove from oven, and place on large serving platter with crackers or fruits dipped in lemon juice to prevent discoloration.

Huckleberry Cheese Rollups

These are delightful little pinwheel-type sandwiches or snacks, and even the children like them as a snack. Not only are they tasty, but they are a healthy alternative to empty calories. These snacks can be made up to 3 days in advance and kept tightly wrapped in the refrigerator.

Ingredients:

      4   oz. cream cheese, room temperature
      4   oz. Feta cheese, crumbled
      ¼   c. dried huckleberries, chopped (recipe page 238)
      ⅓   c. fresh spinach leaves, finely chopped

¼ c. pecans, finely chopped
2   flour tortillas

Directions:

1. Bring cream cheese to room temperature; blend both cheeses together until creamy.
2. Stir in huckleberries, spinach, and pecans.
3. Spread half the mixture on one tortilla; roll tightly and wrap with plastic wrap.
4. Repeat with second tortilla; chill for at least 3 hours.
5. When ready to serve, slice into rounds and place on tray before serving.

Spicy Huckleberry Cheese Spread

This is a delicious spread and may be made with reduced sugar preserves to lower the calories for those who are watching their weight.

Ingredients:

12 oz. huckleberry jam or preserves (recipes pages 192, 197)
2   tsp. prepared horseradish
2   Tbs. Dijon-style prepared mustard
8   oz. cream cheese, softened
    ground black pepper to taste

Directions:

1. In medium bowl combine huckleberry preserves, horseradish, mustard, and black pepper; adjust ingredients to desired taste.
2. Cover and chill overnight if possible.
3. When ready to serve, place cream cheese on serving plate, and pour huckleberry mixture over cream cheese.
4. Serve with basket of your favorite crackers.

Yields: 8 to 10 servings.

Brie Huckleberry Wheel

This is a delicious appetizer that may be prepared in advance and popped into the oven 30 minutes before your guests arrive. Serve it with fresh huckleberries for even more flavor.

Ingredients:

  1   wheel Brie cheese (8 oz.)
  3   Tbs. huckleberry preserves (recipe page 197)
  1   egg white
  9   oz. frozen puff pastry sheets, thawed
      fresh huckleberries as desired

Directions:

  1. Preheat oven to 350 degrees F.
  2. Turn wheel of Brie cheese on edge and slice in half so you have 2 circles of cheese.
  3. Spread huckleberry preserves on cut side of both halves of brie.
  4. Cut each circle in half, and stack cut sides together sandwich style, so that preserves are in center of both halves.
  5. Then face halves back together to assemble one round wheel again; wrap entire wheel of Brie with stacked sheets of puff pastry, overlap, and turn pastry seam side down onto greased baking sheet.
  6. Brush puff pastry with egg white, and bake for 30 minutes until pastry is golden brown.
  7. Remove from oven, and serve immediately with additional fresh huckleberries alongside, if desired, and a knife to cut wedges.

Did You Know? . . . .

Did you know that Northwestern Montana produces the best huckleberry picking in the state?

Huckleberry Delights Cookbook

A Collection of Huckleberry Recipes
Cookbook Delights Series Book 6

Beverages

Table of Contents

Chocoberry Milk Chiller

Chocolate and huckleberries blend to form a cool, refreshing drink.

Ingredients:

- 1   c. chocolate milk
- 4   Tbs. chocolate syrup, divided
- 2   Tbs. huckleberry syrup (recipe page 191)
  multicolored sprinkles, if desired
  fresh or frozen huckleberries, if desired

Directions:

1. Stir 2 tablespoons chocolate syrup and 2 tablespoons huckleberry syrup into chocolate milk, and mix thoroughly.
2. Chill mixture in freezer for 5 minutes.
3. While mixture is chilling, dip rims of 2 chilled glasses upside down into remaining chocolate syrup and then sprinkles to coat edges, if desired.
4. Drizzle remaining chocolate syrup on inside and bottom of glasses.
5. Pull mixture out of freezer; pour into chocolate drizzled glasses.
6. Garnish with huckleberries, if desired.

Yields: 2 servings.

Huckleberry Granita

This makes an excellent-flavored and very refreshing drink. Enjoy!

Ingredients:

- 2   c. fresh huckleberries
- ½   c. sugar
- ½   c. water
  juice from ½ lemon

Directions:

1. Rinse huckleberries and make sure they are free of stems.
2. Blend huckleberries in food processor until completely smooth.
3. Sieve into bowl to get rid of any seeds or skins that did not get processed.
4. Stir in sugar, water, and lemon juice.
5. Put in plastic container and place in freezer for 3 hours, stirring several times until mixture is made up of large crystals but is not completely frozen.
6. After last stirring, spoon into chilled glasses for an icy treat.

Yields: 2 servings.

Appleberry Yogurt Milkshake

This makes an easy apple and berry dairy shake for those who prefer yogurt to ice cream.

Ingredients:

1½ c. frozen vanilla yogurt
⅓ c. sugar
¼ c. honey
½ c. frozen apple juice concentrate, undiluted, thawed
3½ c. milk
½ c. frozen huckleberries

Directions:

1. In blender combine frozen yogurt, sugar, honey, and apple juice concentrate; blend well.
2. Place in container or tray and freeze until almost solid.
3. Remove from freezer and return to blender, adding milk and huckleberries; blend just until smooth.
4. Pour into chilled glasses and serve immediately.

Yields: 6 to 8 servings.

Huckleberry Apple Slush

Looking for a different beverage? Try this combination of apples blended with huckleberries for a new taste twist.

Ingredients:

> 2   c. apple juice, chilled
> 1   c. frozen huckleberries
> ½   c. sugar or to taste
> 12 ice cubes

Directions:

1. Blend apple juice and berries together in blender for 40 seconds.
2. Add sugar and then ice cubes, 2 at a time; cover and blend until smooth.
3. Pour into tall, chilled glasses and serve immediately.

Yields: 4 servings.

Huckleberry Apple Cider Tea

This makes a warm, spicy, and fragrant tea combination for tea drinkers as well as apple cider and huckleberry fans.

Ingredients:

> 8   c. apple cider
> 2   c. water
> 8   Tbs. thick huckleberry syrup (recipe page 189)
> ¼   c. brown sugar, firmly packed
> 1   cinnamon stick
> 1   tsp. whole cloves
> ½   tsp. whole allspice
> 12 tea bags

Directions:

1. In 3-quart saucepan combine apple cider, water, huckleberry syrup, brown sugar, cinnamon, cloves, and allspice.
2. Place over medium heat until mixture starts to boil.
3. Reduce heat and simmer 10 minutes.
4. Remove from heat and add tea bags to hot mixture; cover and let steep 5 minutes.
5. Pour through strainer to remove spices and tea bags.
6. Serve while hot in warmed mugs.

Yields: 10 to 12 servings.

Huckleberry Banana Malted

This is a fruity, flavorful health drink that even the children will enjoy. Why not try one today?

Ingredients:

1   whole ripe banana, peeled
½   c. frozen huckleberries
2   scoops vanilla ice cream
2   c. milk
2   tsp. dry malted milk powder

Directions:

1. Place banana, huckleberries, ice cream, and milk in blender; blend until well mixed.
2. Add malt powder, and blend for 30 seconds more.

Yields: 4 servings.

Huckleberry Breakfast Shake

This makes a tasty, colorful, and nutritious shake to be enjoyed at breakfast or snack time.

Ingredients:

- 2 c. plain or vanilla yogurt
- 1 c. huckleberries, fresh or frozen
- 1 med. ripe banana
- ½ c. orange juice
  small handful granola mix

Directions:

1. Combine yogurt, huckleberries, banana, and orange juice in blender.
2. Blend until smooth and frothy.
3. Pour into chilled glasses and top with granola mix.

Yields: 2 to 4 servings.

Huckleberry Iced Spiced Tea

This makes a refreshing iced tea on a hot day with the addition of cinnamon and ginger. Serve it with cinnamon sticks for swizzles.

Ingredients:

- 6 c. water
- 12 huckleberry herbal tea bags
- 2 cinnamon sticks (3-in. lengths)
- 1 Tbs. fresh ginger, minced
- 1 c. cranberry juice
  sugar to taste
  crushed ice
  cinnamon sticks for swizzles

Directions:

1. Heat water in large saucepan just to boiling point.
2. Add tea bags, cinnamon sticks, and ginger.
3. Remove from heat, cover, and let steep about 15 minutes.
4. Add juice and sugar to taste.
5. Strain tea into pitcher; cover and chill.
6. Pour tea into glasses of crushed ice, and serve garnished with cinnamon sticks as swizzles.

Yields: 6 to 8 servings.

Huckleberry Café au Lait

Try this recipe for a new variation on a favorite coffee drink.

Ingredients:

⅔ c. hot coffee
2   Tbs. huckleberry syrup (recipe page 191)
⅓ c. milk
    cinnamon, nutmeg, or chocolate powder (optional)

Directions:

1. Pour hot coffee into warmed mug.
2. Stir in huckleberry syrup.
3. Steam and froth milk; add to coffee, leaving layer of foam on top.
4. Sprinkle cinnamon, nutmeg, or chocolate powder on top of foam, if desired.
5. Serve immediately.

Yields: 1 serving.

Did You Know? . . . .

Did you know that the huckleberry is the state fruit of Idaho?

Huckleberry Black Soda

This is an easy-to-make, colorful soda to serve on relaxing summer evenings.

Ingredients:

3 Tbs. huckleberry syrup (recipe page 191)
1 Tbs. blackberry syrup
1 c. sparkling water or club soda
  ice

Directions:

1. Fill 16-ounce glass or shaker jar about ¾ full with ice; add both syrups and sparkling water or soda.
2. Stir or shake and serve.

Yields: 1 serving.

Huckleberry Chocolate Cream Coffee

Serve this drink in front of a fire on a cold winter's eve. It will warm you up deliciously.

Ingredients:

8 Tbs. huckleberry syrup (recipe page 191)
4 Tbs. chocolate syrup
1 c. heavy cream, reserve 4 Tbs.
3 c. hot coffee
4 sprinkles of ground cinnamon
4 pinches of grated orange peel
  Sweetened Whipped Cream (recipe page 167)

Directions:

1. Whip together syrups and cream; reserve 4 tablespoons of cream.

2. Stir reserved cream and huckleberry syrup blend in saucepan over low heat until mixed together.
3. Add coffee gradually while stirring mixture.
4. Pour evenly into 4 warmed mugs; top with whipped cream, a sprinkle of cinnamon, and a pinch of grated orange peel.

Yields: 4 servings.

Purple Party Punch

Looking for a fun, purple-colored punch for your party? Then this could be the punch for you. Try floating raspberry sherbet balls on top for an even more festive drink.

Ingredients:

4 Tbs. huckleberry syrup (recipe page 191)
6 oz. powdered grape drink mix
2 c. hot water
4 c. cold water
1 can frozen cranberry juice concentrate (12 oz.)
2 c. pineapple juice, chilled
2 liters ginger ale, chilled
  raspberry sherbet, if desired

Directions:

1. Stir together syrup, drink mix, and hot water; blend well.
2. Add cold water, cranberry juice concentrate, and pineapple juice.
3. Chill mixture until cold then add chilled ginger ale.
4. Pour into large punch bowl.
5. If desired, float balls of sherbet on top after adding ginger ale.

Yields: 20 servings.

Huckleberry Tropical Fruit Smoothie

If you enjoy a tropical, fruity drink, then you will enjoy this fresh smoothie.

Ingredients:

    1¼ c. fresh pineapple chunks
    1   c. huckleberries
    1   banana, peeled, sliced into ½-in. pieces
    2   c. buttermilk
    2   tsp. honey
    4   ice cubes
    4   fresh mint leaves
         fresh huckleberries for garnish

Directions:

1. Place pineapple, huckleberries, and banana in blender; purée until smooth.
2. Add buttermilk, honey, and ice cubes; blend until creamy and smooth.
3. Pour into 4 tall, chilled glasses.
4. Garnish with mint leaf and fresh huckleberries.

Yields: 4 servings.

Purple Cow

When I was little my cousins used to make brown cows and serve them when I would stay overnight. We always enjoyed them, and now there is the purple cow, which is just as good and just as much fun for the kids.

Ingredients:

    2   scoops vanilla ice cream
    2   Tbs. huckleberry syrup (recipe page 191)

10 oz. lemon-lime flavored soda
1½ oz. whipped cream
    frozen huckleberries or maraschino cherries

Directions:

1. Place ice cream in large tumbler glass, and pour huckleberry syrup over it.
2. Pour lemon-lime soda over all.
3. Garnish with whipped cream and frozen huckleberries or a maraschino cherry on top.
4. Serve with a straw and a long spoon.

Yields: 1 large or 2 small servings.

Wild Huckleberry Smoothie

This is an easy-to-make smoothie that is very refreshing.

Ingredients:

6 oz. wild huckleberries, fresh or frozen
6 oz. vanilla yogurt
1 Tbs. honey
½ c. ice (3 ice cubes)

Directions:

1. Place all ingredients in blender; blend well at high speed.
2. Serve immediately in chilled glasses.

Yields: 1 serving.

Did You Know? . . . .

Did you know that in the Pacific Northwest region of North America, the huckleberry plant is found in mid-alpine regions, frequently on the lower slopes of mountains?

Huckleberry Almond Milk

This is a delightfully refreshing drink that is particularly good on a hot summer day. Not only is it delicious, it is also healthy. Be sure to prepare your almonds a day or two in advance. If you prefer, you may substitute the sugar with honey or another sweetener of choice.

Ingredients:

2	c. unsalted, dry-roasted almonds
1½	c. water
½	c. heavy cream
⅓-½	c. huckleberries, fresh or frozen, to taste
⅓-½	c. sugar to taste

Directions:

1. Place almonds and water in tightly closed jar, and store in refrigerator for 1 to 2 days at most.
2. Pour mixture into blender and blend until smooth.
3. Strain mixture through fine sieve, reserving almond pulp for another use.
4. Note: Use 4 to 6 ice cubes if you prefer your beverage more like a shake, but be sure to reduce water by ½ cup or more. If you use ice cubes, process them with the strained liquid first, then add remaining ingredients.
5. Pour strained liquid back into blender, and add heavy cream, berries, and sugar to taste, blending all until smooth and creamy.
6. Pour into chilled glasses and serve immediately.

Yields: 2 to 4 servings.

Did You Know? . . . .

Did you know that huckleberries ripen in mid to late summer?

Huckleberry Delights Cookbook

A Collection of Huckleberry Recipes
Cookbook Delights Series Book 6

Breads and Rolls

Table of Contents

Page

Applesauce Huckleberry Muffins

This apple, bran, and huckleberry combination makes great bran muffins. They are delicious and moist.

Ingredients:

1½ c. all-purpose flour
⅓ c. sugar
1   Tbs. baking powder
½   tsp. salt
1   tsp. ground cinnamon
1   c. unsweetened applesauce
2   eggs
1   c. butter, softened
¾   c. milk
1¼ c. bran flakes cereal, slightly crushed
1   c. huckleberries, fresh or frozen

Directions:

1. Preheat oven to 400 degrees F.
2. Coat 12-cup muffin tin with nonstick spray.
3. In large mixing bowl sift together flour, sugar, baking powder, salt, and cinnamon.
4. In small bowl cream together applesauce, eggs, and butter; add milk and cereal, blending well.
5. Stir liquid ingredients into dry ingredients until dry ingredients are just moistened, about 20 strokes.
6. Gently fold in huckleberries; spoon batter into oiled muffin tin, dividing evenly.
7. Bake 20 to 25 minutes until muffins are browned or until wooden pick inserted in center comes out clean.
8. Allow muffins to cool for about 2 minutes in pan, then transfer to wire rack to cool.

Yields: 12 muffins.

Berry Best Huckleberry Muffins

Buttermilk and applesauce make this a very moist, tasty muffin.

Ingredients:

- 1 c. all-purpose flour
- ¾ c. whole-wheat flour
- ¾ c. sugar
- 1 Tbs. baking powder
- 1 tsp. lemon peel, finely shredded
- 2 egg whites
- ⅔ c. buttermilk
- ⅓ c. applesauce
- 1 tsp. vanilla extract
- 1 c. huckleberries, fresh or frozen, unthawed

Directions:

1. Preheat oven to 400 degrees F.
2. Spray 6 large (3-inch) or 12 standard muffin cups with nonstick spray.
3. In large bowl stir together both flours, sugar, baking powder, and lemon peel, making sure all are evenly distributed; make well in center of mixture.
4. In small bowl beat egg whites until foamy; stir in buttermilk, applesauce, and vanilla.
5. Add liquid mixture to well in dry mixture, stirring until just moistened; fold in huckleberries.
6. Fill oiled muffin cups ¾ full with batter.
7. Bake for 22 to 25 minutes or until wooden pick inserted in center comes out clean.
8. Cool muffins in pan for 5 minutes; remove and cool on wire rack.
9. Note: For standard-size muffins bake for 18 to 20 minutes or until done.

Yields: 6 large or 12 standard muffins.

Chocolate Huckleberry Bagels

This is a delicious, chewy treat that is wonderful on its own and even more wonderful when split, toasted, and buttered.

Ingredients:

- 5 c. unbleached all-purpose flour
- 2 pkg. active dry yeast (.25 oz. each)
- ¼ c. sugar
- 3 Tbs. salt
- 1 tsp. ground cinnamon
- 4 tsp. baking powder
- 2 tsp. baking soda
- 3 lg. eggs
- 1 c. vegetable oil
- 1 c. warm water (110 degrees F.)
- 1 c. semisweet chocolate chips
- 1 c. fresh huckleberries

Directions:

1. Combine flour, yeast, sugar, salt, cinnamon, baking powder, and soda.
2. In separate bowl combine eggs, oil, and warm water; add to dry mixture to form wet dough.
3. Gently stir in chocolate chips and huckleberries.
4. Turn out onto floured surface; divide into 12 pieces.
5. Roll each piece into long snake, then form into shape of a bagel.
6. Place bagels on baking sheet and bake for about 1 hour.
7. Remove from oven and cool on wire rack.

Yields: 12 bagels.

Did You Know? . . . .

Did you know that you should freeze huckleberries within a few days of picking them?

Huckleberry Mandarin Orange Muffins

The pecans and mandarin oranges add a fruity, nutty flavor to these delightfully fresh-tasting muffins.

Ingredients:

6	c. cake flour
2	Tbs. plus 1 tsp. baking powder
1	Tbs. baking soda
¼	tsp. salt
1½	c. brown sugar, firmly packed
¾	c. sugar
3	eggs
3	c. plain yogurt
1	Tbs. orange or vanilla extract
½	c. canola oil
3	c. mandarin orange segments, drained, halved
1	c. pecans, chopped
4	c. huckleberries

Directions:

1. Preheat oven to 400 degrees F.
2. Oil muffin tins.
3. In large bowl combine flour, baking powder, baking soda, and salt; set aside.
4. In another bowl combine sugars, eggs, yogurt, extract, and oil; stir into dry mixture and mix just to combine.
5. Gently fold in oranges, pecans, and huckleberries.
6. Scoop ¼ cup batter into each prepared muffin tin.
7. Bake for 18 to 22 minutes or until firm to the touch.

Yields: 36 muffins.

Jamaican Jungle Bread

This is a flavorful, fruit-filled bread recipe originating in Jamaica. Try it next time you want a healthy snack.

Ingredients:

1½ c. all-purpose flour
1   c. quick-cooking oats
½   c. sugar
1   tsp. baking powder
1   tsp. baking soda
1   tsp. ground cinnamon
½   c. flaked coconut
½   c. toasted nuts, chopped
½   c. mashed bananas
½   c. mashed huckleberries
8   oz. crushed pineapple, undrained
¼   c. vegetable oil
2   lg. eggs

Directions:

1. Preheat oven to 350 degrees F.
2. Butter and flour loaf pan.
3. In large bowl stir together flour, oats, sugar, baking powder, soda, cinnamon, coconut, and nuts.
4. In medium bowl mix bananas, huckleberries, pineapple, oil, and eggs.
5. Add all at once to dry mixture; stir just until moistened.
6. Spoon into prepared loaf pan.
7. Bake for 50 minutes or until wooden pick inserted in center comes out clean.
8. Cool in pan for 5 minutes; remove from pan and cool completely on wire rack.

Yields: 1 loaf.

Huckleberry Bran Muffins

These are great for breakfast on a cold winter morning or for a Sunday brunch.

Ingredients:

1	c. bran cereal flakes
½	c. skim milk
1	c. unsweetened applesauce
⅓	c. molasses
⅓	c. sugar
2	Tbs. vegetable oil
1	egg
1	egg white
1½	c. all-purpose flour
2½	tsp. baking powder
½	tsp. salt
½	tsp. ground cinnamon
1	c. huckleberries, fresh or frozen, unthawed

Directions:

1. Preheat oven to 400 degrees F.
2. Butter 12-cup muffin tin.
3. In large bowl combine bran cereal and milk; let stand 5 minutes.
4. Add applesauce, molasses, sugar, oil, egg, and egg white; stir until well combined.
5. In medium bowl stir together flour, baking powder, salt, and cinnamon; add to wet ingredients and stir just until moistened.
6. Gently fold in huckleberries.
7. Spoon equally into buttered muffin cups, and bake for 20 to 25 minutes.
8. Remove from oven; tip muffins out of pan onto wire rack.
9. Serve warm or cold as desired.

Yields: 12 muffins.

Huckleberry Cream Cheese Muffins

These muffins are rich and flavorful. They are so good they do not need any condiments and are best when fresh and hot from the oven.

Ingredients:

- 2 c. cake flour
- ¾ c. sugar
- 1½ tsp. baking powder
- ½ tsp. baking soda
- 3 oz. cream cheese, softened
- 2 tsp. lemon juice
- 2 tsp. vanilla extract
- 2 eggs
- 4 Tbs. hot, melted butter
- ½ c. milk
- 1 c. huckleberries

Directions:

1. Preheat oven to 350 degrees F.
2. Line cups of muffin tin with 12 paper liners.
3. Combine flour, sugar, baking powder, and soda in large mixing bowl.
4. In another bowl using mixer, blend cream cheese, lemon juice, and vanilla until smooth.
5. Add eggs and blend 30 seconds; scrape down sides of bowl with spatula.
6. With mixer running, pour hot, melted butter into mixture, and continue blending another 30 seconds.
7. Add milk and blend 10 seconds.
8. Stir in dry ingredients by hand just until mixed.
9. Gently fold in huckleberries, and mix until all are incorporated into batter.
10. Pour equal amount of batter into each muffin cup, filling each about ¾ full.
11. Bake for 20 to 25 minutes, until top springs back when touched with finger.
12. Remove from oven; turn out onto wire rack to cool slightly before serving.

Yields: 12 muffins.

Little Blue House Nut Bread

This is great nut bread, and with the addition of huckleberries, it will have a nice purple color. Try it at your next brunch or afternoon tea.

Ingredients:

- 3 c. all-purpose flour
- 1 Tbs. baking powder
- ¼ tsp. baking soda
- ½ tsp. ground nutmeg
- ¾ c. sugar
- 3 lg. eggs
- 2 tsp. vanilla extract
- ½ c. milk
- ½ c. butter, melted
- ½ Tbs. orange extract
- ⅓ c. orange juice
- 3 c. huckleberries, fresh or frozen
- 1¼ c. chopped walnuts or pecans

Directions:

1. Preheat oven to 350 degrees F.
2. Butter 4 mini loaf pans or 2 regular loaf pans; set aside.
3. Combine flour, baking powder, soda, nutmeg, and sugar.
4. Using wooden spoon, beat together eggs, vanilla, milk, melted butter, orange extract, and orange juice.
5. Add to dry ingredients, and stir until flour is thoroughly moistened.
6. Gently fold in huckleberries and nuts.
7. Pour into prepared loaf pans, and set aside for 15 minutes.
8. Bake about 45 minutes for mini loaves (about 75 minutes for large loaves) or until wooden pick inserted into center comes out clean.
9. Remove from oven and let sit in pans for 8 minutes; turn out onto wire racks to cool completely before slicing to serve.

Yields: 4 mini loaves or 2 large loaves.

Huckleberry Yam Bread

This is a delicious, slightly dense bread that will pack well for sack lunches or snacks. You may use canned or fresh-cooked yams, sweet potatoes, or even canned pumpkin.

Ingredients:

    2   lg. eggs, slightly beaten
    1⅓ c. sugar
    ⅓   c. canola oil
    1   c. mashed yams or sweet potatoes
    1   tsp. vanilla extract
    1½ c. all-purpose flour (plus extra for dusting pan)
    1   tsp. ground cinnamon
    ½   tsp. ground allspice
    1   tsp. baking soda
    1   c. huckleberries

Directions:

1. Preheat oven to 350 degrees F.
2. Grease and flour 9 x 5-inch loaf pan.
3. In large bowl combine eggs, sugar, oil, yams, and vanilla; blend well.
4. In separate bowl combine flour, cinnamon, allspice, and baking soda.
5. Make well in center and pour yam mixture into well; mix just until moistened.
6. Gently fold in huckleberries; spoon batter into prepared loaf pan.
7. Bake for 1 hour or until wooden pick inserted in center comes out clean.
8. Cool 20 minutes in pan, then turn out onto wire rack to cool completely before slicing to serve.

Yields: 1 large loaf.

Huckleberry Zucchini Bread

This is a great way to use up some of the extra zucchini you may have. With the addition of huckleberries, even the children enjoy this zucchini bread.

Ingredients:

- 3 lg. eggs, lightly beaten
- 1 c. vegetable oil
- 3 tsp. vanilla extract
- 2¼ c. sugar
- 3 c. all-purpose flour
- 1 tsp. salt
- 1 tsp. baking powder
- ¼ tsp. baking soda
- 1 Tbs. ground cinnamon
- 2 c. shredded zucchini
- 2 c. fresh huckleberries

Directions:

1. Preheat oven to 350 degrees F.
2. Grease and flour 4 mini loaf pans.
3. In large bowl beat together eggs, oil, vanilla, and sugar.
4. Beat in flour, salt, baking powder, baking soda, and cinnamon.
5. Gently fold in zucchini and huckleberries.
6. Spoon batter into prepared mini loaf pans.
7. Bake for 50 minutes or until knife inserted in center of loaf comes out clean.
8. Cool 20 minutes in pans, then turn out onto wire racks to cool completely before slicing to serve.

Yields: 4 mini loaves.

Streusel Topped Huckleberry Bread

This bread is so delicious it tastes almost like a dessert. It is great with tea or coffee and goes well with milk for the kids!

Ingredients for bread:

2   c. plus 2 Tbs. all-purpose flour, divided
2   tsp. baking powder
½   tsp. salt
½   c. butter, softened
¾   c. sugar
2   lg. eggs
1   tsp. vanilla extract
¼   tsp. lemon zest
½   c. milk
1½ c. fresh huckleberries

Ingredients for streusel:

2   Tbs. all-purpose flour
5   Tbs. sugar
½   tsp. ground cinnamon
2   Tbs. butter, diced

Directions for bread:

1. Preheat oven to 375 degrees F.
2. Butter 13 x 9 x 2-inch baking pan.
3. In medium bowl combine 2 cups flour, baking powder, and salt.
4. In large bowl beat butter with sugar until light and fluffy; beat in eggs, vanilla, and lemon zest.
5. Stir in dry ingredients alternately with milk.
6. Toss berries with 2 tablespoons flour; fold into batter.
7. Spoon batter into prepared baking pan.

Directions for streusel:

1. Combine flour, sugar, and cinnamon in small bowl.

2. Cut in cold, diced butter with fork or pastry blender until mixture resembles coarse crumbs; sprinkle over batter in pan.
3. Bake for 30 to 35 minutes or until wooden pick inserted in center of bread comes out clean.
4. Cool in pan on wire rack before cutting into squares for serving.

Yields: 16 to 18 servings.

Frosty Huckleberry Fritters

These fritters are very good, but they are best when hot from the fryer. Vary your topping by mixing cinnamon in with the powdered sugar.

Ingredients:

1½ c. all-purpose flour
¾ c. sugar
1   Tbs. baking powder
¾ c. fresh huckleberries
1   egg, beaten
1   c. milk
    powdered sugar
    ground cinnamon, if desired

Directions:

1. Preheat 2 inches of oil in deep skillet to 375 degrees F.
2. Sift together flour, sugar, and baking powder; stir in berries.
3. Mix together egg and milk; add to dry mixture, blending to completely moisten.
4. Drop batter by tablespoonfuls into hot oil; fry until golden brown, 3 to 4 minutes, turning once.
5. Drain and cool to warm on paper towels.
6. Sprinkle with powdered sugar; mix with cinnamon if desired for variation.

Wild Huckleberry Scones

What a wonderful way to serve these wild huckleberry scones—warm from the oven with butter and honey, or for a traditional flavor, serve them with cream and jam.

Ingredients:

>   2   c. all-purpose flour
>   1   Tbs. baking powder
>   3   Tbs. sugar
>   ½   tsp. ground allspice
>   5   Tbs. butter, softened
>   ½   c. milk
>   ½   c. huckleberries, fresh or frozen
>   milk for glaze
>   sugar for glaze

Directions:

1. Preheat oven to 425 degrees F.
2. Combine flour, baking powder, sugar, and allspice in large bowl.
3. Add butter, cutting in with pastry knife or fork until in small, crumb-size pieces.
4. Make well in center and add milk; stir several strokes.
5. Fold in huckleberries and stir just until dough forms; add a sprinkle more of milk if dough does not stick together, but do not overwork dough.
6. Turn out onto lightly floured surface; form into ball and flatten into 9- or 10-inch disk.
7. Using sharp, long-bladed knife, cut dough into 8 or 10 wedges, as you would cut a pie.
8. Either slide dough onto buttered baking sheet as is for scones shaped in wedges, or separate pieces and push them into slightly flattened balls for scones shaped in circles.
9. For either shape, brush tops with milk.
10. For sweet, crunchy surface, sprinkle sugar on each one.

11. Bake for 12 to 15 minutes or until golden brown; may take a bit longer if baked as a big disk that has been scored into wedges but not separated.
12. Remove from baking sheet when done; cut wedges apart at existing score marks.
13. Serve while hot in cloth-lined basket, and enjoy these unique scones with family or friends!

Yields: 8 to 10 scones.

Buttermilk Huckleberry Muffins

Adding buttermilk to these muffins makes them a tasty treat.

Ingredients:

1¾ c. all-purpose flour
⅔   c. sugar
1   Tbs. baking powder
¼   tsp. salt
1   tsp. lemon gelatin mix
¾   c. buttermilk
1   egg, lightly beaten
¼   c. butter, melted
1   c. huckleberries, fresh or frozen, unthawed

Directions:

1. Preheat oven to 400 degrees F.
2. Line muffin cups with paper liners.
3. Mix dry ingredients together in large bowl.
4. Combine buttermilk, egg, and butter; add to dry ingredients and stir just until moist.
5. Gently fold in huckleberries.
6. Fill paper-lined muffin cups ¾ full.
7. Bake for about 20 minutes or until wooden pick inserted in center comes out clean.
8. Cool in pan for 5 minutes, then remove to wire rack to cool.

Yields: 12 muffins.

Huckleberry Bread

This tasty bread is great for a quick snack. Make plenty, as it will go fast with the kids or the man of the house!

Ingredients:

1¾ c. sugar
3 c. all-purpose flour, sifted
1½ tsp. baking soda
1 Tbs. ground cinnamon
¾ tsp. salt
1¼ c. oil
4 eggs, beaten
2 c. huckleberries, fresh or frozen

Directions:

1. Preheat oven to 350 degrees F.
2. Oil and lightly flour bread pan.
3. Mix sugar, flour, baking soda, cinnamon, and salt together in large mixing bowl.
4. Add oil and eggs; stir just until mixed.
5. Gently fold in huckleberries.
6. Pour into prepared pan, and bake for 1 hour or until wooden pick inserted in center comes out clean.
7. Remove from oven, let sit in pan for 8 minutes, then turn out onto wire rack to cool completely before slicing.

Yields: 1 loaf.

Did You Know? . . . .

Did you know that the "garden huckleberry" is not a true huckleberry but is a member of the nightshade family, which includes not only nightshade but tomatoes and chili peppers?

Huckleberry Delights Cookbook

A Collection of Huckleberry Recipes
Cookbook Delights Series Book 6

Breakfasts

Table of Contents

Page

Cottage Cheese Huckleberry Pancakes

These pancakes are hearty with the addition of cottage cheese and even more enjoyable with the addition of fresh huckleberries for a tasty treat.

Ingredients:

- 2 c. huckleberries, fresh or frozen, divided
- ½ c. sugar
- ¼ tsp. ground cinnamon
- ¼ c. butter, melted
- 6 eggs
- 1 c. cottage cheese
- 2 Tbs. vegetable oil
- 1 tsp. vanilla extract
- ½ c. all-purpose flour
- 1 tsp. baking powder
- Sweetened Whipped Cream (recipe page 167)

Directions:

1. In saucepan mix all but ¼ cup huckleberries with sugar, cinnamon, and butter; cook over moderate heat until berries are soft.
2. Cover and keep warm until ready to serve.
3. With mixer blend eggs, cottage cheese, oil, and vanilla.
4. Combine flour and baking powder; add to egg mixture and blend well.
5. Pour ¼ cup for each pancake on greased griddle or in frying pan.
6. Cook on both sides until done; place on plate and keep warm until all batter is used.
7. Place cakes on individual serving plates, and spoon about 2 tablespoons of huckleberry mixture on one side of each pancake; fold other side over.
8. Top with whipped cream and garnish with remaining huckleberries.

Yields: 12 pancakes.

French Pancakes with Fresh Huckleberries and Whipped Cream

French pancakes are very similar to crepes and, yes, both are a family favorite. These are served without syrup but with fresh huckleberries and whipped cream instead.

Ingredients:

- 4 eggs, beaten well
- 2 c. milk
- ½ tsp. salt
- 2 c. all-purpose flour
  fresh huckleberries
  Sweetened Whipped Cream (recipe page 167)

Directions:

1. In bowl beat together eggs, milk, salt, and flour until smooth.
2. Cover and let stand 30 minutes.
3. Batter should be thin, just thick enough to coat a spoon dipped in it.
4. If batter is too thick, stir in a little more milk.
5. Heat 5- or 6-inch frying pan, and grease lightly with vegetable oil.
6. Pour in just enough batter to cover pan with very thin layer.
7. Tilt and turn pan so batter spreads evenly.
8. If there is a little too much, tip pan over mixing bowl, and pour extra back.
9. Cook on one side, turn with spatula, and brown other side.
10. Place on plate to keep warm, and repeat with remaining batter.
11. When ready to serve, roll up or fold in quarters and place on individual plates.
12. Top each serving with whipped cream and fresh berries.

French-Style Croissants with Huckleberry Sauce

Try something different and use croissants instead of bread for your French toast.

Ingredients for sauce:

1½ c. frozen huckleberries
¼ c. sugar
½ c. water
½ c. huckleberry jam (recipe page 192)
Sweetened Whipped Cream (recipe page 167)

Ingredients for fried croissant:

3 Tbs. butter
2 eggs, beaten
2 Tbs. milk
4 croissants, cut in half

Directions for sauce:

1. Combine huckleberries, sugar, water, and huckleberry jam in saucepan.
2. Simmer over medium heat for approximately 20 to 30 minutes or just until thickened, stirring occasionally.
3. Remove from heat, and keep warm while preparing croissants.
4. This sauce may be stored in refrigerator and also used to top ice cream, pound cake, pancakes, waffles, or French toast.

Directions for fried croissant:

1. Heat butter to bubbling in frying pan or on griddle.
2. Whisk eggs and milk together with fork.
3. Dip both sides of bottom half of croissant and only inside of top half in egg mixture.
4. Brown sides that have egg mixture on them in frying pan until golden.
5. Turn bottom only once, cooking until browned; do not turn top half, leaving it to stay dry and attractive.

6. Place bottom half of cooked croissant on plate.
7. Top with 2 tablespoons warm huckleberry sauce.
8. Place top half of croissant slightly askew over bottom half.
9. Use 2 more tablespoons of huckleberry sauce to drizzle over top half.
10. Top with whipped cream and serve immediately.

Yields: 4 servings.

Huckleberry Corn Pancakes

The hearty taste of corn blends well with the fruity flavor of huckleberries for a winning flavor combination in these pancakes.

Ingredients:

1½ c. yellow cornmeal
1 tsp. baking soda
¼ c. whole-wheat flour
1 tsp. salt
2 Tbs. honey
3 Tbs. oil
2 c. buttermilk
1 egg, lightly beaten
1½ c. huckleberries

Directions:

1. Combine cornmeal, baking soda, flour, and salt in medium bowl; set aside.
2. Combine honey, oil, buttermilk, and egg; beat well to blend.
3. Stir this mixture quickly into dry ingredients.
4. Let stand for 10 minutes to soften cornmeal.
5. Gently fold in huckleberries.
6. Lightly grease preheated skillet; allow about ¼ cup batter for each pancake, and pour onto bottom of heated skillet.
7. Cook until bubbly; turn once and cook on other side until golden brown.
8. Place on individual plates and serve hot with butter and syrup.

Huckleberry Breakfast Cake

This makes a great-tasting breakfast cake. Use either yogurt or sour cream as an attractive addition for a company breakfast or brunch.

Ingredients:

3   eggs, beaten
1½ c. milk
¾   c. all-purpose flour
⅓   c. sugar
3   Tbs. butter, melted
8   oz. plain yogurt or sour cream
¼   c. brown sugar, firmly packed
3   c. huckleberries, sugared to taste
1   kiwi fruit for garnish

Directions:

1. Preheat oven to 375 degrees F.
2. With blender mix eggs, milk, flour, and sugar.
3. Pour melted butter into fluted quiche pan.
4. Pour egg mixture into pan, and bake for 30 minutes or until edges are brown and center is set.
5. Remove from oven and let cool for 5 minutes.
6. Turn over onto serving tray, remove pan, and let cool completely.
7. Mix yogurt or sour cream with brown sugar; fill center with mixture.
8. Top with sugared huckleberries and garnish with sliced kiwi fruit; serve immediately.

***Did You Know?* . . . .**

Did you know that the Native Americans used to use the branches of the huckleberry plant as brooms?

Huckleberry Breakfast Custard

This is an interesting combination of cheese and huckleberries in an egg custard. Serve with toast or your favorite muffins.

Ingredients:

> 5 eggs, beaten
> 1¼ c. milk
> 3 Tbs. butter, melted
> 1 tsp. cornstarch
> ⅛ tsp. baking powder
> ¼ tsp. salt
> ¼ tsp. pepper
> ¼ tsp. dried mustard
> ⅔ c. cheddar cheese, shredded
> nonstick vegetable spray
> fresh huckleberries

Directions:

1. Preheat oven to 425 degrees F.
2. Beat together eggs, milk, butter, cornstarch, baking powder, salt, pepper, and mustard.
3. Divide cheese among 4 individual custard dishes that have been sprayed with nonstick spray.
4. Pour mixture evenly into custard dishes.
5. Place custard dishes into baking pan that is at least 2 inches deep; fill pan with boiling water 1 inch deep.
6. Bake uncovered for 15 to 20 minutes.
7. Remove from oven, place on individual serving dishes, and spoon fresh huckleberries over each.

Yields: 4 servings.

Did You Know? . . . .

Did you know that the most likely purple fruit to fool a beginner huckleberry picker is the serviceberry?

Huckleberry Buttermilk Pancakes

These buttermilk pancakes are full of huckleberries, are easy to make, and will disappear quickly.

Ingredients:

2   lg. eggs
2   Tbs. sugar
1   c. buttermilk
1   tsp. vanilla extract
4   Tbs. butter, melted, cooled, divided
¾   c. unsifted all-purpose flour
½   tsp. baking soda
2   c. fresh huckleberries
     powdered sugar
     warm huckleberry syrup (recipe page 191)

Directions:

1. In small bowl lightly beat eggs together with sugar; stir in buttermilk, vanilla, and 2 tablespoons melted butter.
2. In large bowl mix together flour and baking soda; add liquid ingredients, and stir until just moistened.
3. Batter should have consistency of thick cream with some lumps; do not overmix.
4. Heat large griddle or 2 large nonstick skillets over medium-high heat; brush each lightly with some of remaining melted butter.
5. Gently drop batter onto skillet by heaping tablespoonfuls, 2 inches apart.
6. Press a few huckleberries into each pancake and cook until undersides are golden brown and bubbles are breaking on top, about 1½ minutes; turn and cook another 1½ minutes.
7. Keep pancakes warm in oven set at low temperature as you cook remaining batter.
8. Divide pancakes among warmed plates and top with remaining huckleberries.
9. Sprinkle with powdered sugar and serve with warm huckleberry or maple syrup.

Huckleberry Skillet Soufflé

This is a nice soufflé and makes a great breakfast dish.

Ingredients:

    4 c. fresh huckleberries, divided
    ⅓ c. huckleberry preserves (recipe page 197)
    1 Tbs. framboise (raspberry-flavored liqueur)
    2 Tbs. butter
    8 lg. egg whites
    ¼ tsp. cream of tartar
    ¼ c. plus 2 Tbs. sugar, divided

Directions:

1. Preheat oven to 375 degrees F.
2. Mash 1 cup huckleberries and the preserves with potato masher in bowl.
3. Stir in framboise.
4. Start to melt butter in deep, 10- or 11-inch skillet over low heat.
5. Meanwhile, beat egg whites and cream of tartar to soft peaks in large mixing bowl.
6. Gradually beat in ¼ cup sugar; beat to stiff peaks.
7. Fold huckleberry mixture into whites ⅓ at a time, just until blended.
8. Increase heat to medium-low.
9. Scrape mixture into skillet, gently spreading it to sides and mounding in center.
10. Cook over medium-low heat for 2 minutes without stirring.
11. Transfer skillet to oven and bake soufflé 15 minutes, until set.
12. Remove from oven and top with remaining berries tossed with 2 tablespoons sugar.

Yields: 4 servings.

Huckleberry French Toast

One of my daughters loves French toast, and this makes an enjoyable variation of the classic version.

Ingredients for sauce:

2 c. frozen huckleberries
¾ c. maple syrup
1 tsp. grated orange peel
1 Tbs. cornstarch
2 Tbs. water

Ingredients for toast:

4 eggs, beaten
¾ c. milk
1 tsp. vanilla extract
¼ tsp. ground nutmeg
8 slices bread
4 Tbs. butter
   powdered sugar

Directions for sauce:

1. Combine huckleberries, maple syrup, and orange peel in small saucepan.
2. Dissolve cornstarch in water; add to huckleberry mixture.
3. Cook and stir until mixture boils; reduce heat and simmer 1 minute or until mixture thickens.
4. Cover and keep warm while preparing toast.

Directions for toast:

1. Beat together eggs, milk, vanilla, and nutmeg; pour into shallow bowl.
2. Dip each slice of bread individually into egg mixture just before placing in lightly buttered skillet or on buttered grill.
3. Cook each slice about 2 minutes or until golden brown.

4. Place one slice on each individual plate; spread with 3 tablespoons huckleberry sauce.
5. Sprinkle lightly with powdered sugar and serve.

Yields: 8 servings.

Huckleberry Crepes with Ricotta Cheese

Our family loves crepes, and the combination of huckleberries and ricotta cheese in this recipe makes a great filling. Leftover crepes are also very good reheated the next day.

Ingredients:

  2 c. all-purpose flour
  6 lg. eggs
  ½ tsp. salt
  3 c. milk
  1 c. ricotta cheese
  2 c. fresh huckleberries
     powdered sugar
     mint leaves

Directions:

1. In large bowl mix together flour, eggs, salt, and milk until smooth.
2. Heat 8-inch crepe pan; brush with butter.
3. Pour ¼ cup batter into pan.
4. Tip and cover entire pan with batter.
5. Crepe is ready to turn when bottom side begins to brown.
6. Slip cooked crepes onto warm plate, and keep warm while remaining batter is used.
7. When ready to serve, spread 2 tablespoons ricotta cheese down middle of crepe, place huckleberries on top, and roll crepe together.
8. Place on individual plates and sprinkle with powdered sugar; garnish with mint leaves and huckleberries.
9. Serve plain or with pure maple syrup.

Huckleberry Puffs

Our family enjoys cream puffs and French donuts, so these were a natural. This is a great treat to enjoy for an occasional breakfast.

Ingredients:

- 1   c. water
- 1   pinch of salt
- ½   c. butter
- 1   c. all-purpose flour
- 4   lg. eggs
- ¾   c. fresh huckleberries
- ½   c. powdered sugar
-     oil for frying

Directions:

1. Preheat oil in deep skillet to 350 degrees F.
2. In medium saucepan bring water, salt, and butter to a boil, completely melting butter.
3. Remove pan from heat, and vigorously whisk in flour until mixture forms ball.
4. Immediately whisk in eggs one at a time.
5. Fold in huckleberries.
6. Drop small balls of dough into hot oil; cook until batter puffs and expands and turns golden brown.
7. Remove huckleberry puffs from oil, and set on paper towels to drain; cover to keep warm.
8. Repeat for remaining batter.
9. Sprinkle with powdered sugar and serve while warm.

***Did You Know?* . . . .**

Did you know that researchers in Idaho who have been trying for years to produce commercial huckleberry plants now hope to have the first ones available in about 2010?

Sour Cream Soufflé with Huckleberries

Try this interesting soufflé. The Parmesan cheese and sour cream reduce the sweetness and are a great contrast to the sweet huckleberry.

Ingredients:

- 6   lg. egg yolks
- ½   c. sour cream
- ¼   c. Parmesan cheese, grated
- 6   egg whites, stiffly beaten
- 3   Tbs. butter
- ½   c. sour cream sweetened with l tsp. sugar or to taste
  fresh huckleberries

Directions:

1. Preheat oven to 325 degrees F.
2. Beat egg yolks until thick and pale colored, about 5 minutes.
3. Mix together sour cream and Parmesan cheese; add ½ cup egg yolks.
4. Fold in stiffly beaten egg whites.
5. Melt butter in 10-inch cast iron skillet.
6. Scrape in egg mixture, leveling gently.
7. Cook over very low heat for 10 minutes, uncovered.
8. Carefully move to oven and bake for 15 minutes until golden and puffed.
9. Serve in pan at the table; cut into 4 wedges.
10. Top each serving with dollop of sweetened sour cream and top with fresh berries.

Yields: 4 servings.

Did You Know? . . . .

Did you know that the name "huckleberry" is used by many people to identify wild blueberries?

Pecan Waffles with Huckleberry Sauce

The combination of huckleberries and pecans adds to the flavor of these waffles. In addition, the warm huckleberry sauce with a touch of lemon makes a great topping.

Ingredients for waffles:

4	egg yolks
2	c. buttermilk
⅔	c. butter
1⅓	tsp. vanilla extract
1⅓	c. sour cream
2⅔	c. all-purpose flour
1⅓	tsp. baking powder
1⅓	tsp. baking soda
⅓	c. sugar
4	egg whites
¾	c. pecans, chopped

Ingredients for sauce:

½	c. sugar
4	Tbs. cornstarch
2	c. apple juice
¼	c. butter
1	tsp. grated lemon rind
1	c. huckleberries, fresh or frozen

Directions for waffles:

1. In large bowl mix together egg yolks, buttermilk, butter, vanilla, and sour cream.
2. In separate bowl combine flour, baking powder, soda, and sugar; add to liquid ingredients, mixing just until blended.
3. Beat egg whites and fold into mixture; add pecans.
4. Spray waffle iron once with nonstick cooking spray, and allow to heat up before adding batter.
5. Cook waffles following manufacturer's instructions.

Directions for sauce:

1. Combine sugar and cornstarch in saucepan; add juice, butter, lemon rind, and huckleberries.
2. Cook over medium heat until sauce thickens.
3. Remove from heat and cool slightly; serve while warm over waffles on individual plates.

Huckleberry Waffles

Huckleberries are a delicious addition to waffles. Serve them hot with homemade huckleberry syrup.

Ingredients:

1½ c. all-purpose flour
1 Tbs. baking powder
½ tsp. salt
3 egg yolks, beaten
1½ c. buttermilk
6 Tbs. butter, melted
3 egg whites
2 Tbs. brown sugar
1 c. huckleberries

Directions:

1. Grease waffle irons well; heat to very hot.
2. Sift together flour, baking powder, and salt.
3. Combine beaten egg yolks with buttermilk and melted butter; beat in flour mixture.
4. Beat egg whites until stiff; add sugar and beat again.
5. Fold into batter, then gently stir in huckleberries.
6. Cook waffles in hot waffle iron, following manufacturer's instructions.
7. Serve with huckleberry topping and/or whipped cream.

Cheese Blintzes with Huckleberries

These crepes make excellent blintzes. Huckleberry preserves and fresh huckleberries make great toppings.

Ingredients for crepes:

2	lg. eggs
2	Tbs. vegetable oil
1	c. milk
¾	c. all-purpose flour, sifted

Ingredients for filling and toppings:

½	c. cream cheese, softened
2	egg yolks
½	c. cottage cheese
2	Tbs. sugar
1	tsp. vanilla extract
	huckleberry preserves (recipe page 197)
	fresh huckleberries
	sour cream
	butter for frying

Directions for crepes:

1. Beat eggs, oil, and milk together.
2. Add flour slowly, blending until smooth.
3. Chill 30 minutes. (Batter should be consistency of heavy cream. If too thick, add a little milk.)
4. Grease hot, 8-inch skillet with light coating of butter.
5. Pour 3 to 4 tablespoons of batter into skillet, tipping and turning pan to coat it.
6. Fry lightly on one side then remove from skillet.
7. Repeat with remaining batter.
8. Stack crepes on wax paper, browned side up.

Directions for filling and toppings:

1. Beat together cream cheese, egg yolks, cottage cheese, sugar, and vanilla until smooth.
2. Fill each crepe with 2 heaping tablespoons filling; fold in sides of crepe over filling and roll up as for egg rolls.

3. Crepes may be frozen at this stage and fried later without defrosting.
4. Melt butter over medium heat in large skillet; add rolled crepes seam side down, and fry until golden brown on all sides.
5. Remove from pan, drain on paper towel, and then place on serving dish.
6. Top with dollop of sour cream and/or huckleberry preserves and fresh huckleberries.

Huckleberry Oven Custard Puff Pancake

Our family loves puff pancakes. The addition of huckleberries makes this a great breakfast treat.

Ingredients:

2	Tbs. butter
3	eggs
2	c. milk, divided
¾	c. all-purpose flour
1	Tbs. sugar
½	tsp. salt
2	c. huckleberries
¼	tsp. ground cinnamon

Directions:

1. Preheat oven to 425 degrees F.
2. Heat butter in heavy skillet in oven until bubbly.
3. While skillet and butter are heating, beat eggs, ¼ cup milk, flour, sugar, and salt together in bowl until smooth; beat in remaining milk.
4. Pour into oven-heated skillet; bake for 20 minutes.
5. Remove from oven and sprinkle with berries and cinnamon.
6. Bake 10 to 15 minutes longer, until knife inserted comes out clean and pancake is browned and puffed.
7. Serve immediately in pan at the table.
8. Cut into wedges to serve; sprinkle with additional fresh huckleberries.

Sour Cream Huckleberry Waffles

Try these great-tasting waffles for a nice change of pace and a delicious breakfast with the family.

Ingredients for waffles:

    6   lg. eggs
    3   c. sour cream
    3   c. all-purpose flour
    9   Tbs. sugar
    1½ tsp. baking soda
    15 Tbs. melted butter

Ingredients for sauce:

    1½ c. frozen huckleberries, divided
    ½   c. orange juice, divided
    1   Tbs. cornstarch
        sugar to taste

Directions for waffles:

1. In large bowl beat eggs with whisk until light; fold in sour cream.
2. Sift together flour, sugar, and baking soda; stir into egg mixture; add melted butter and blend lightly.
3. Spray waffle iron with nonstick cooking spray once before cooking; use 1 cup mix for 4-section waffle iron.
4. Cook waffles according to the manufacturer's instructions.

Directions for sauce:

1. Combine ¾ cup huckleberries, ¼ cup orange juice, and sugar to taste in saucepan over low heat until warm.
2. Mix together remaining ¼ cup orange juice with cornstarch; add to saucepan and cook until mixture has thickened.
3. Remove from heat; add remaining ¾ cup huckleberries, folding in gently to keep berries whole.
4. Cover and keep warm until ready to serve.
5. Spoon over waffles and serve immediately.

Yields: 12 waffles.

Huckleberry Delights Cookbook

A Collection of Huckleberry Recipes
Cookbook Delights Series Book 6

Cakes

Table of Contents

Page

Crumb Topped Huckleberry Oatmeal Cake

This is a great cake for those times when you have a lot of people to serve. It can be served warm out of the oven, so there is no need to wait for cooling or frosting it. Just serve with butter and leave room for seconds.

Ingredients for cake:

- 10 Tbs. butter
- 1½ c. quick-cooking oats
- 2 c. boiling water
- ½ c. sugar
- 1 c. brown sugar, firmly packed
- ½ tsp. salt
- 1½ tsp. baking soda
- 2 c. all-purpose flour
- 1½ tsp. ground cinnamon
- 3 lg. eggs, beaten
- 1 c. fresh huckleberries

Ingredients for crumb topping:

- ⅔ c. sugar
- ⅔ c. all-purpose flour
- 1 tsp. ground cinnamon
- ¼ c. nuts, chopped
- ½ c. butter

Directions for cake:

1. Preheat oven to 350 degrees F.
2. Butter and flour 13 x 9 x 2-inch baking pan.
3. Cut butter into chunks, and mix with oats and boiling water in heatproof bowl; cover and let stand for 5 minutes.
4. Mix together sugar, brown sugar, salt, baking soda, flour, and cinnamon.
5. Stir oat mixture into flour mixture; add beaten eggs and blend well.
6. Gently fold in fresh huckleberries, and spoon batter into prepared pan.

Directions for crumb topping:

1. Mix together sugar, flour, cinnamon, and nuts.
2. Cut in butter with pastry cutter or two knives until mixture resembles crumbs about size of peas.
3. Sprinkle over top of batter in pan.
4. Bake for 40 to 45 minutes or until cake tests done.
5. Remove from oven and cool to lukewarm in pan on wire rack before cutting into squares to serve.

Yields: 12 to 14 servings.

Cheesecake with Huckleberries

This is an easy-to-make version of cheesecake that is so tasty they will be back for seconds!

Ingredients:

2   c. cream cheese, softened
¾   c. sugar
¼   tsp. vanilla extract
2   lg. eggs
1   prepared graham cracker crust (recipe page 217)
1½   c. fresh huckleberries, divided

Directions:

1. Preheat oven to 350 degrees F.
2. Beat together cream cheese, sugar, and vanilla until smooth and creamy.
3. Add eggs and blend well.
4. Pour into graham cracker pie crust.
5. Spoon ½ cup huckleberries over top; gently swirl with wooden pick.
6. Bake for 40 minutes or until center is set.
7. Remove from oven and cool to room temperature; refrigerate to chill completely.
8. Remove from refrigerator when ready to serve; slice and top with remaining fresh huckleberries.

Yields: 8 servings.

Fresh Huckleberry Teacakes

These delicious little teacakes are best made in mini muffin tins. Serve warm out of the oven with butter, or drizzle a glaze over them if preferred.

Ingredients:

- 2 c. all-purpose flour
- ¾ tsp. baking soda
- ¾ tsp. ground cinnamon
- ¼ tsp. ground nutmeg
- ¼ tsp. ground ginger
- 1 c. sugar
- 2 lg. eggs at room temperature
- 2 tsp. vanilla extract
- ½ c. vegetable oil
- 4 Tbs. melted butter
- 1 c. fresh huckleberries

Directions:

1. Preheat oven to 400 degrees F.
2. Butter mini muffin tins.
3. Blend together flour, baking soda, cinnamon, nutmeg, and ginger in large mixing bowl; set aside.
4. Whisk together sugar and eggs in small bowl until light, about 2 minutes.
5. Blend in vanilla, oil, and melted butter.
6. Make large well in center of dry ingredients, and pour in liquid mixture; combine quickly until batter is formed.
7. Very gently, fold in huckleberries.
8. Spoon batter into prepared muffin tins, filling them ⅔ full.
9. Bake about 20 minutes or until firm to the touch and wooden pick inserted in center comes out clean.
10. Remove from oven, let stand for 5 minutes, and then invert pan onto wire rack, tipping out muffins to cool.
11. Serve either warm or cooled with butter or glaze if desired.

Yields: 12 to 16 teacakes.

Huckleberries and Sour Cream Cake

This is a simple, hearty cake similar to a pound cake, which my husband always enjoys. Try it glazed or with a cream cheese frosting.

Ingredients:

- ½ c. butter, softened
- 1 c. sugar
- 3 lg. eggs
- 1 tsp. vanilla extract
- 1 c. sour cream
- 2 c. sifted all-purpose flour
- 1 tsp. baking soda
- 1¼ c. huckleberries
- ½ c. brown sugar, firmly packed
- ¾ c. hazelnuts, chopped
- ½ tsp. ground cinnamon

Directions:

1. Preheat oven to 350 degrees F.
2. Grease 13 x 9 × 2-inch baking pan.
3. In large bowl cream together butter and sugar; beat in eggs one at a time, beating until smooth.
4. Add vanilla and stir in sour cream; beat well.
5. Combine flour and baking soda; mix into creamed mixture.
6. Gently fold in huckleberries; spread half of batter into prepared pan.
7. In bowl mix brown sugar, nuts, and cinnamon together, blending completely.
8. Sprinkle mixture over top of batter; spread remaining batter over top.
9. Bake 40 minutes or until wooden pick inserted in center comes out clean.
10. Remove from oven and cool on wire rack.

Yields: 24 servings.

Huckleberry Cake

This cake is truly decadent. The combination of huckleberries, pears, apricots, and apples adds so much flavor and moisture that it does not need any frosting. It also packs well, making it a great snack for lunch box treats.

Ingredients:

- ½ c. butter, softened
- 1 c. sugar
- 3 lg. eggs
- 2¼ c. all-purpose flour, divided
- 1 Tbs. baking powder
- ½ tsp. salt
- 1 tsp. grated orange peel
- ½ c. milk
- 2 c. frozen huckleberries, partially thawed
- ½ c. fresh pears, pared, cored, diced
- ½ c. apples, pared, cored, diced
- ½ c. dried apricots, chopped

Directions:

1. Preheat oven to 350 degrees F.
2. Butter and lightly flour 10 x 4-inch tube pan.
3. Cream butter and sugar until light and fluffy; beat in eggs one at a time, beating well after each addition.
4. Combine 2 cups flour, baking powder, salt, and orange peel, mixing well.
5. Add flour mixture and milk alternately to creamed mixture.
6. Combine fruits; toss with remaining flour to lightly coat.
7. Spread half of batter in prepared tube pan.
8. Spoon fruit over batter, and top with remaining batter.
9. Bake for about 1 hour or until golden brown and wooden pick inserted near center comes out clean.
10. Remove from oven; cool in pan on wire rack for 10 minutes.
11. Place serving plate over top of cake in pan and carefully turn over; remove pan from cake.
12. Cut into wedges and serve warm or cold.
13. To marinate flavors, cake can be wrapped tightly in plastic wrap for storing after completely cooled.

Huckleberry Cream Cheese Coffee Cake

This is a rich and tasty coffee cake, great to serve for brunch or when having afternoon tea.

Ingredients:

- ¼ c. butter, softened
- 1 c. cream cheese, softened
- 1 c. plus 2 Tbs. sugar, divided
- 1 lg. egg
- 1 tsp. vanilla extract
- 1 c. all-purpose flour
- 1 tsp. baking powder
- 2 c. fresh huckleberries
- 1 tsp. ground cinnamon
-   vegetable cooking spray

Directions:

1. Preheat oven to 350 degrees F.
2. Coat 9-inch round cake pan with cooking spray.
3. In large mixing bowl cream together butter and cream cheese; gradually add 1 cup sugar and blend well.
4. Add egg and vanilla; beat well.
5. Combine flour and baking powder; stir into creamed mixture, then gently fold in huckleberries.
6. Pour batter into prepared cake pan.
7. Combine cinnamon with remaining 2 tablespoons sugar; sprinkle over top of batter.
8. Bake for 1 hour or until wooden pick inserted in center comes out clean.
9. Remove from oven and let cool before serving.

Yields: 8 servings.

Huckleberry Italian Cheesecake

The addition of huckleberries makes this a very tasty variation of the classic Italian-style cheesecake.

Ingredients for crust:

    1¼ c. graham cracker crumbs
    3   Tbs. sugar
    ⅓   c. butter, melted

Ingredients for filling:

    5   lg. eggs
    1   c. sugar
    3   Tbs. all-purpose flour
    ¼   tsp. ground nutmeg
    ½   tsp. ground mace
    30 oz. whole-milk ricotta cheese
    1   pt. fresh huckleberries

Directions for crust:

1. In medium bowl using fork, combine graham cracker crumbs and sugar; blend in melted butter.
2. Press evenly on bottom of 9-inch springform pan.

Directions for filling:

1. Preheat oven to 350 degrees F.
2. In large mixing bowl beat eggs until light and fluffy.
3. Gradually add sugar, flour, nutmeg, and mace, beating until well blended.
4. Add ricotta cheese and mix well.
5. With rubber spatula, fold in huckleberries.
6. Pour into crust and place pan on cookie sheet.
7. Bake for 75 minutes or until center is almost set.
8. Turn off heat and leave oven door closed; let cake sit in oven an additional 20 minutes.
9. Remove cake from oven and cool in pan on wire rack.
10. After cake is cooled, place in refrigerator and chill 4 or more hours.

11. When ready to serve, run knife around edge of pan to loosen cake; remove sides of pan.
12. Garnish top with fresh huckleberries, whipped cream, and mint leaves, if desired.

Yields: 8 to 10 servings.

Huckleberry Honey Cake

This is a cake made in a loaf pan and sliced like bread. You may drizzle a simple glaze over the top if desired.

Ingredients:

½   c. butter, softened
½   c. sugar
½   c. honey
3   lg. eggs, beaten
½   c. milk
1½ c. plus 1 Tbs. unbleached all-purpose flour, divided
2   tsp. baking powder
⅛   tsp. salt
1   c. huckleberries, fresh or frozen, well drained

Directions:

1. Preheat oven to 350 degrees F.
2. Grease and flour 9 x 5-inch loaf pan.
3. In large mixing bowl cream together butter, sugar, and honey.
4. Beat in eggs and milk; blend well.
5. Sift together 1½ cups flour, baking powder, and salt; add to creamed ingredients and blend thoroughly.
6. In small bowl toss huckleberries with 1 tablespoon flour; gently fold berries into batter.
7. Pour batter into prepared loaf pan.
8. Bake for about 1 hour, until cake is golden brown and wooden pick inserted in center comes out clean.
9. Remove from oven, and cool in pan 10 minutes; turn out onto wire rack to cool before slicing to serve.

Yields: 1 large loaf.

Huckleberry Lemon Pound Cake

For a real treat, top a slice of this sumptuous pound cake with a generous scoop of huckleberry sorbet.

Ingredients:

1	c. unsalted butter, softened
1	c. sugar
4	lg. eggs, separated
1	Tbs. lemon zest, finely grated
1	tsp. vanilla extract
¾	c. sour cream
1½	c. all-purpose flour
1	tsp. baking powder
1	pinch of salt
1¾	c. huckleberries, lightly rinsed, drained

Directions:

1. Preheat oven to 350 degrees F.
2. Butter loaf pan and line bottom with parchment or wax paper; butter paper.
3. Lightly dust pan with flour, shaking out excess.
4. In mixing bowl cream butter and sugar with electric mixer until light and fluffy.
5. Add egg yolks, lemon zest, and vanilla extract, beating until smooth.
6. Beat in sour cream, blending well.
7. Sift flour, baking powder, and salt together; add to creamed mixture in three batches stirring with wooden spoon just until combined.
8. Beat egg whites in bowl until soft peaks form.
9. Fold whites into batter ⅓ at a time to combine.
10. Gently fold in huckleberries and scrape batter into prepared pan.
11. Bake for 35 minutes.
12. Reduce oven temperature to 325 degrees F., and bake an additional 45 minutes or until golden brown and wooden pick inserted in center comes out clean.

13. Cool cake in pan on wire rack for at least 1 hour.
14. Remove cake from pan, place back on rack, and remove paper lining.
15. Allow cake to cool completely before slicing to serve.

Yields: 1 loaf.

Huckleberry Pound Cake

My husband loves huckleberries and pound cake. This is an excellent pound cake with the added treat of huckleberries.

Ingredients:

1	c. butter, softened
2	c. sugar
4	eggs
1	tsp. vanilla extract
3	c. all-purpose flour, divided
¼	tsp. salt
1	tsp. baking powder
2	c. huckleberries

Directions:

1. Preheat oven to 325 degrees F.
2. Butter 10-inch tube or bundt pan and coat with sugar.
3. In large bowl cream butter and sugar.
4. Add eggs one at a time, beating well until light and fluffy; add vanilla.
5. Sift 2 cups flour, salt, and baking powder together; add to creamed mixture and beat well.
6. Dredge berries in remaining flour; gently fold into creamed mixture.
7. Spoon into prepared pan; bake for 1 hour and 15 minutes.
8. Remove from oven; let cool in pan for 8 minutes, and then turn out onto wire rack to cool before slicing to serve.

Yields: 10 to 12 servings.

Huckleberry Shortcake

Most of us are used to strawberry shortcake, but this huckleberry shortcake makes a flavorful and colorful change to satisfy the palate.

Ingredients for biscuits:

    1¾ c. all-purpose flour
    3   Tbs. baking powder
    ¼   c. sugar
    4   Tbs. butter
    ¾   c. milk

Ingredients for topping:

    2   c. fresh huckleberries
    ¾   c. sugar
    6   scoops vanilla ice cream
        Sweetened Whipped Cream (recipe page 167)
        whole fresh huckleberries

Directions for biscuits:

1. Preheat oven to 425 degrees F.
2. Combine flour, baking powder, and sugar in medium mixing bowl; cut in butter with pastry knife until mixture resembles small peas.
3. Quickly stir in milk, and mix with wooden spoon for 30 seconds.
4. Spread onto well-floured board and knead gently for 15 seconds.
5. Roll to ½-inch thickness, and cut into 3-inch circles with round cookie cutter or jar cap; place on baking sheet.
6. Bake for 12 to 20 minutes or until golden.

Directions for topping and assembly:

1. Combine huckleberries and sugar in blender; blend until thick purée forms.
2. On 6 individual plates pour 3 tablespoons huckleberry purée.
3. On top of purée, place bottom half of biscuits that have been sliced in half horizontally.

4. Place whole fresh huckleberries and whipped cream on top of each biscuit; add scoop of ice cream.
5. Top with other half of biscuit and garnish over all with more purée, huckleberries, and sweetened whipped cream in an attractive manner.
6. Serve immediately.

Yields: 6 servings.

Wild Huckleberry Cake

What a wonderful and delicious way to treat your family to those berries you all just picked! This is just scrumptious with French vanilla ice cream.

Ingredients:

1¾ c. plus 1 Tbs. all-purpose flour, divided
1   c. sugar
1   tsp. ground cinnamon
¾   c. butter
1   tsp. baking soda
½   c. milk
2   eggs, slightly beaten
1   c. wild huckleberries

Directions:

1. Preheat oven to 325 degrees F.
2. Butter 9 x 12-inch baking pan.
3. In large bowl sift together 1¾ cup flour, sugar, and cinnamon; cut in butter.
4. Dissolve soda in milk; add eggs and beat well.
5. Add milk mixture to dry ingredients; blend well.
6. Lightly sprinkle huckleberries with 1 tablespoon flour to keep them from sinking in batter.
7. Gently fold into batter, just until mixed in.
8. Pour into prepared baking pan, and bake for 1 hour or until wooden pick inserted in center comes out clean.
9. Remove from oven, and place on wire rack to cool before slicing to serve.
10. Serve with whipped cream or try French vanilla ice cream on the side.

Huckleberry Spice Cake

Huckleberries have a very sharp taste and hold up well to the molasses and spices in this cake. This is a great variation for spice cake fanatics.

Ingredients:

    2   c. all-purpose flour, sifted
    2   tsp. baking powder
    1   tsp. baking soda
    ½   tsp. ground cinnamon
    ½   tsp. ground cloves
    ½   tsp. ground allspice
    ⅓   c. butter, softened
    1   c. sugar
    1   egg, at room temperature, beaten
    3   Tbs. molasses
    1   c. sour milk or buttermilk
    1   pt. huckleberries
    ⅓   c. powdered sugar

Directions:

1. Preheat oven to 375 degrees F.
2. Butter and lightly flour 13 x 9 x 2-inch baking pan.
3. Sift together flour, baking powder, baking soda, and spices.
4. In large mixing bowl cream butter with sugar.
5. Add beaten egg, and cream together until lemon colored; gradually beat in molasses.
6. Add this mixture, alternating with milk, to dry ingredients, a little at a time, beating well after each addition.
7. Sprinkle berries with a little flour; gently fold into batter.
8. Pour into prepared baking pan; bake for about 30 minutes or until wooden pick inserted in center comes out clean.
9. Remove from oven, and cool cake in pan on wire rack.
10. Dust with powdered sugar; cut into squares to serve.

Yields: 12 to 14 servings.

Streusel Topped Huckleberry Cake

This makes a moist cake with the crunchy and delicious taste of streusel topping.

Ingredients for cake:

3	Tbs. butter, softened
1	c. sugar
1	egg
1¾	c. sifted all-purpose flour
2	tsp. baking powder
1	c. milk
1½	c. fresh huckleberries, rinsed, drained, patted dry

Ingredients for crumb topping:

½	c. all-purpose flour
½	c. sugar
2	Tbs. butter, softened
1	tsp. ground cinnamon

Directions for cake:

1. Preheat oven to 350 degrees F.
2. Butter 11 × 7-inch baking pan.
3. Cream butter with sugar; beat in egg.
4. Sift together flour and baking powder; add to creamed mixture alternately with milk.
5. Sprinkle huckleberries with a little flour to keep from sinking; stir into batter.
6. Pour batter into prepared baking pan.

Directions for crumb topping:

1. Mix ingredients together until crumbly; sprinkle over top of batter.
2. Bake until cake tests done, about 40 minutes.
3. Remove from oven, and cool on wire rack before slicing to serve.

Huckleberry Comfort Shortcake

This is a simple, colorful dessert to serve using homemade pound cake.

Ingredients for cake:

2½ c. sugar
1   c. butter, softened
1   tsp. vanilla or almond extract
5   lg. eggs
3   c. all-purpose flour, plus extra for dusting pan
1   tsp. baking powder
¼   tsp. salt
1   c. milk or evaporated milk
     powdered sugar

Ingredients for topping:

2½ c. frozen sliced peaches, thawed
2   c. huckleberries
½   c. sugar
⅓   c. Southern Comfort whiskey (optional), for flavor
     Sweetened Whipped Cream (recipe page 167)

Directions for cake:

1. Preheat oven to 350 degrees F.
2. Grease and lightly flour bottom and sides of 10 × 4-inch tube pan.
3. Beat sugar, butter, vanilla, and eggs in large bowl with electric mixer on low speed for 30 seconds, scraping bowl occasionally.
4. Mix flour, baking powder, and salt.
5. On low speed beat flour mixture into wet mixture alternately with milk, beating just until smooth after each addition; pour into prepared pan.
6. Bake for 70 to 80 minutes or until wooden pick inserted in center comes out clean.
7. Remove from oven, and cool 20 minutes in pan on wire rack; turn over onto serving plate.
8. Cool completely, about 2 hours; sprinkle with powdered sugar.
9. Slice cake and assemble with topping per directions below.

Directions for topping and assembly:

1. In large bowl combine peach slices and huckleberries.
2. Sprinkle sugar over fruit mixture and stir in Southern Comfort or flavoring of choice; let stand about an hour.
3. Spoon fruit mixture over slices of pound cake, and top with sweetened whipped cream.

Yields: 10 servings.

Huckleberry Frosting

Use this tasty huckleberry filling and frosting on your favorite white layer cake.

Ingredients:

2	pkg. cream cheese (8 oz. each), softened
1	c. unsalted butter, softened
6¼	c. powdered sugar
½	c. fresh huckleberries, puréed
2	tsp. grated lemon peel
1½	tsp. vanilla extract
1½	pt. fresh huckleberries

Directions:

1. Beat cream cheese and butter in mixing bowl.
2. Gradually beat in powdered sugar.
3. Add huckleberry purée, peel, and vanilla.
4. Chill frosting for about 30 minutes until firm but still spreadable.
5. Place one cake layer on serving platter.
6. Place generous amount of frosting on layer, creating a ridge around outside edge.
7. Place 1 cup fresh huckleberries on top of frosting.
8. Dot huckleberries with additional frosting.
9. Place second layer over first, and repeat frosting process.
10. Top with last layer, and frost entire cake with thin layer of frosting; refrigerate for 30 minutes.
11. Add a more generous layer of frosting.
12. Top center of cake with more huckleberries.

Yields: Enough for three 9-inch layers.

Huckleberry Banana Snack Cake

This is one of our children's favorite after-school snacks, and we like to give them this delicious, nutritious treat.

Ingredients:

1¼ c. sugar
⅔ c. butter, melted
¼ c. buttermilk
2 lg. eggs
1 tsp. vanilla extract
2 c. ripe bananas, mashed
2 c. all-purpose flour
¼ tsp. baking soda
⅛ tsp. salt
1 c. huckleberries, frozen or fresh
powdered sugar

Directions:

1. Preheat oven to 350 degrees F.
2. Grease 9 × 9-inch baking pan.
3. In large bowl beat together sugar, butter, buttermilk, eggs, and vanilla until smooth.
4. Add bananas; continue beating until creamy and well mixed.
5. Add flour, baking soda, and salt; beat at low speed just until all ingredients are moistened.
6. Using spatula, gently fold in huckleberries.
7. Spoon batter into prepared baking pan.
8. Bake 25 to 30 minutes or until wooden pick inserted in center comes out clean.
9. Remove from oven and cool on wire rack in pan.
10. When ready to serve, sprinkle top with powdered sugar, then cut into squares.

Yields: 16 servings.

Huckleberry Delights Cookbook

A Collection of Huckleberry Recipes
Cookbook Delights Series Book 6

Candies

Table of Contents

Page

Easy Chocolate and Huckleberry Truffles

These are some rich, delicious, melt-in-your-mouth candies. Make lots, as they do go fast!

Ingredients:

8   oz. semisweet chocolate, coarsely chopped
4   oz. unsweetened chocolate
8   Tbs. unsalted butter
1   can sweetened condensed milk (14 oz.)
6   Tbs. orange-flavored liqueur
½   tsp. orange zest, finely grated
½   c. huckleberry fruit leather (recipe page 240) or dried berries, finely chopped (recipe page 238)
1   c. almonds, toasted, finely chopped
½   c. instant chocolate drink powder

Directions:

1. Heat both chocolates, butter, and milk in saucepan until chocolates and butter are partially melted.
2. Remove from heat and continue to stir until completely melted.
3. Whisk in flavoring and orange zest until creamy and smooth; stir in huckleberry bits.
4. Transfer to bowl, and let stand until firm enough to hold its shape, about 2 hours.
5. Using a tablespoon, mold chocolate into balls, 1 level tablespoon at a time; place on cookie sheet lined with buttered parchment paper.
6. Place chopped almonds and chocolate powder in small bowl; mix well.
7. Working with one at a time, drop truffles into bowl with greased fingertips.
8. Shake bowl back and forth so truffles are completely coated.
9. Return to parchment paper to set for about 1 hour before serving.

10. Can be refrigerated in airtight container up to 5 days or frozen up to 1 month before serving; let stand at room temperature to soften slightly.

Yields: 3 dozen balls.

Chocolate Covered Huckleberries

What a delicious and wonderful way to eat your fruit and have some chocolate, too!

Ingredients:

1   c. semisweet chocolate chips
1   Tbs. unsalted butter
2   c. fresh huckleberries, rinsed, dried

Directions:

1. Melt chocolate in glass bowl in microwave or in metal bowl set over pan of simmering water.
2. Stir frequently until melted and smooth; remove from heat and stir in butter until melted.
3. Cool mixture somewhat; add huckleberries to chocolate, stirring gently to coat.
4. Quickly spoon small clumps of candy mixture onto wax paper-lined baking sheet.
5. Refrigerate until firm, about 10 minutes, before serving.
6. May be stored in cool place in airtight container for 2 to 3 days.
7. Note: These also may be made with chopped huckleberry fruit leather in place of fresh berries. They will store up to 1 month in airtight container.

Yields: 2½ dozen pieces.

Huckleberry Caramels

Homemade caramels are delicious. Try this easy recipe, and you will want to make them often. Other flavors of syrup and dried fruit can be substituted if desired.

Ingredients:

- 1   c. sugar
- ½   c. dark corn syrup
- ¼   c. huckleberry syrup (recipe page 191)
- ½   c. butter
- 1   c. light cream, divided
- 1   c. nuts, chopped
- ⅓   c. dried huckleberries (recipe page 238)
- 1   tsp. vanilla extract

Directions:

1. Combine sugar, syrups, butter, and ½ cup cream in saucepan.
2. Bring to boil over medium heat, stirring constantly.
3. Add remaining cream; continue stirring, cooking slowly to very hard ball stage (268 degrees F.).
4. Remove from heat, and add nuts, dried berries, and vanilla; pour into buttered 8- or 9-inch square pan.
5. When cooled completely and firm, cut into squares of size desired.
6. Wrap each square with piece of plastic wrap or wax paper for storing.
7. Store in airtight container for up to 4 weeks or in freezer up to 3 months.

Did You Know? . . . .

Did you know that huckleberries have pink flowers, and blueberries typically have white flowers?

Huckleberry Cheesecake Fudge

This is a fun variation on traditional fudge and is very rich and creamy.

Ingredients:

⅔ c. evaporated milk
2½ c. sugar
5 oz. marshmallow cream
¼ c. butter
3 oz. cream cheese, room temperature
12 oz. white chocolate chips
1½ c. dried huckleberries (recipe page 238)
1 tsp. vanilla extract

Directions:

1. Heat milk over medium heat until warm; add sugar.
2. Over medium heat bring mixture to rolling boil, stirring constantly with wooden spoon.
3. Remove from heat; add marshmallow cream and butter.
4. Bring back to rolling boil for 5½ minutes by the clock; start timing once rolling boil resumes.
5. Cut cream cheese into small pieces to allow easy melting, and add to boiling mixture about 1 minute before end of boil.
6. If brown flakes appear in mixture, turn down heat a little.
7. Remove from heat, and add white chocolate chips and huckleberries.
8. Stir until creamy and all chips are melted; stir in vanilla extract.
9. Mix thoroughly and pour into lightly buttered 9 x 9-inch baking pan.
10. Cool on wire rack, then cut into squares before serving.
11. May be stored in airtight container in refrigerator for up to 3 weeks or in freezer up to 3 months.

Yields: 16 to 20 pieces.

Huckleberry Divinity

These are wonderful, easy-to-make treats that are enjoyable for all. The purple juice adds an attractive color to the divinity.

Ingredients:

> 2   c. sugar
> ½   c. huckleberry juice (recipe page 242)
> ½   tsp. salt
> 1   pt. marshmallow cream
> ½   c. nuts
> 1   tsp. vanilla extract

Directions:

1. In medium saucepan boil together sugar, juice, and salt until it forms hard ball in cool water.
2. Place marshmallow cream in mixing bowl; beat in hot syrup.
3. Continue beating until slightly stiff and will hold a peak.
4. Fold in nuts and vanilla.
5. Drop from spoon onto wax paper; allow to sit until firm.
6. May be kept in airtight container for up to 3 weeks or freeze for up to 3 months.

Huckleberry Fudge Balls

These are very tasty and will disappear quickly once your family finds them.

Ingredients:

> 8   oz. cream cheese, softened
> 6   oz. semisweet chocolate morsels, melted
> ¾   c. vanilla wafer crumbs
> ¼   c. huckleberry preserves (recipe page 197)
> ¾   c. almonds, toasted, finely chopped

Directions:

1. In medium bowl beat cream cheese with electric mixer on medium speed until creamy.
2. Add melted chocolate, beating until smooth.
3. Stir in vanilla wafer crumbs and preserves; cover and chill until of soft, firm consistency.
4. Shape into 1-inch balls; roll in chopped almonds.
5. Place on wax paper-covered baking sheet; chill until firm.
6. May be stored in airtight container up to 3 weeks or in freezer up to 3 months.

Huckleberry Cream Fudge

Huckleberry cream fudge is a great treat here in the Northwest. Enjoy this colorful, creamy fudge.

Ingredients:

2 c. sugar
1 c. light cream
1 Tbs. light corn syrup
½ tsp. salt
1 Tbs. butter
2 tsp. vanilla extract
¾ c. dried huckleberries (recipe page 238)

Directions:

1. In heavy, 2-quart saucepan combine sugar, cream, corn syrup, and salt.
2. Bring to boil over moderate heat, stirring constantly.
3. Cook to soft-ball stage (238 degrees F.).
4. Remove from heat, and cool to 110 degrees F. or to lukewarm; do not stir.
5. Add butter and vanilla; beat until mixture becomes very thick and loses its gloss.
6. Stir in huckleberries and spread in 8 × 8-inch buttered pan.
7. Score squares while fudge is still warm.
8. When cooled, cut into squares to serve.

Huckleberry Nut Candy with Chocolate

Here is a delicious and chewy chocolate candy with the zing of huckleberry flavor.

Ingredients:

- ½ c. butter
- 1 can sweetened condensed milk (14 oz.)
- 1 pkg. chocolate chips (12 oz.)
- 1 pkg. butterscotch chips (6 oz.)
- ½ c. pecans, chopped fine
- ½ c. huckleberry fruit leather, chopped fine (recipe page 240)
- 1 pkg. miniature marshmallows (10.5 oz.)
- 1 tsp. vanilla extract

Directions:

1. In saucepan combine butter and condensed milk; bring to a boil then remove from heat.
2. Stir in chocolate chips and butterscotch chips until melted.
3. Stir in nuts, huckleberry bits, marshmallows, and vanilla.
4. Drop by spoonfuls onto wax paper.
5. Let sit until firm, then they may be wrapped or stored in airtight container in refrigerator for up to 3 weeks or freeze for up to 3 months.

Yields: 2 to 2½ dozen pieces.

Did You Know? . . . .

Did you know that the Huckleberry Railroad, a heritage train located in Flint, Michigan, was named that because it ran so slow a person could jump off the train, pick huckleberries, and jump back on the train with little effort?

Huckleberry Nut Fudge

My children love this fudge, and it is a breeze to make. Be sure to make plenty, because they can eat it up almost as fast as you make it!

Ingredients:

- 4 c. brown sugar, firmly packed
- 8 oz. evaporated milk
- 3 Tbs. butter
- ⅓ c. corn syrup
- 2 Tbs. huckleberry juice (recipe page 242)
- ¾ c. walnuts, chopped
- ½ c. dried huckleberries (recipe page 238)

Directions:

1. Over medium-high heat, bring sugar, milk, butter, syrup, and huckleberry juice to full boil; boil exactly 4 minutes.
2. Remove from heat, and beat with electric mixer for 10 minutes; stir in nuts and huckleberries.
3. Spread into greased 8 x 8-inch pan; cut into squares before it cools completely.
4. Turn out onto board, wrap individually if desired, and store in airtight container.
5. This fudge also freezes well for up to 3 months.

Did You Know? . . . .

Did you know that while some Vaccinium *species, such as the Red Huckleberry, are always called huckleberries, other species may be called blueberries or huckleberries depending upon local custom, and that similar* Vaccinium *species in Europe are called bilberries?*

Huckleberry Nuggets

Fruit and nut nugget candy rolled in powdered sugar is a very popular candy in the Northwest. This is one of my daughters' favorite candies.

Ingredients:

2   env. unflavored gelatin
1½ c. huckleberry purée, divided
2   c. sugar
1   c. walnuts, chopped
¼   tsp. almond extract
     powdered sugar

Directions:

1. Sprinkle gelatin over ½ cup of huckleberry purée; set aside and let stand.
2. Combine remaining huckleberry purée and sugar in saucepan; bring to boil over moderate heat, stirring constantly.
3. Add softened gelatin mixture and continue stirring; bring to rolling boil and cook for 15 to 20 minutes longer.
4. Stir in nuts and almond extract.
5. Pour into buttered 8 x 8-inch pan.
6. Cool to room temperature; refrigerate overnight.
7. Cut into 1 x 1-inch pieces.
8. Carefully invert onto firm surface to remove from pan; roll each piece in powdered sugar.
9. Allow to stand 2 to 3 days before serving.

Did You Know? . . . .

Did you know that seven species of huckleberry have been identified in the state of Montana?

Huckleberry Gumdrops

These delicious old-fashioned gumdrops can be made right in your own home where the children can have fun helping.

Ingredients:

1	pkg. powdered fruit pectin (1¾ oz.)
¾	c. water
½	c. baking soda
1	c. sugar
1	c. light corn syrup
2	Tbs. huckleberry syrup (recipe page 191)
	sugar for coating

Directions:

1. Combine pectin, water, and baking soda in medium saucepan. (This mixture will foam.)
2. Combine sugar and corn syrup in large saucepan.
3. Place both pans on stove over high heat.
4. Cook, stirring alternately, until foam disappears from pectin mixture and sugar mixture boils rapidly, about 8 minutes.
5. Pour pectin mixture into boiling sugar mixture in thin stream, until all pectin mixture is added.
6. Then, stirring constantly, boil 1 minute longer.
7. Remove from heat.
8. Stir in huckleberry syrup; immediately pour mixture into 8 x 8 x 2-inch pan.
9. Let stand at room temperature for 3 hours or until candy is cool and firm. (Do not refrigerate.)
10. Cut gumdrops into cubes with knife dipped in warm water.
11. Roll cubes in sugar and let stand at room temperature 1 hour; repeat rolling and standing.
12. Store in airtight container.

Huckleberry Surprises

What a delicious surprise is in store with these scrumptious candies.

Ingredients:

> ½   c. butter, softened
> 1¾ c. powdered sugar
> 1   tsp. orange juice
> 1½ c. shredded coconut
> 30 pieces huckleberry fruit leather, cut in ½-in. squares (recipe page 240)
>   powdered sugar

Directions:

1. In medium bowl cream together butter, sugar, and orange juice; mix in coconut.
2. Fold fruit leather pieces in half; wrap coconut mixture around each piece to cover completely, and roll in powdered sugar.
3. Store in refrigerator in tightly covered container until ready to serve.
4. May be kept in freezer for up to 3 months in airtight container.

Yields: 2½ dozen pieces.

Did You Know? . . . .

Did you know that there are dark-colored blueberries and distinctly blue huckleberries?

Did you know that the way to tell blueberries and huckleberries apart is that blueberries have many tiny, soft seeds, but huckleberries have ten large, bony seeds?

Huckleberry Delights Cookbook

A Collection of Huckleberry Recipes
Cookbook Delights Series Book 6

Cookies

Table of Contents

Amaretto Huckleberry Butter Cookies

These are buttery cookies, decorated with chopped almonds and huckleberry bits, and flavored with orange peel and almond-flavored liqueur.

Ingredients:

    1   c. butter, room temperature
    1   c. sugar
    1   lg. egg, separated
    3   Tbs. almond-flavored liqueur, such as amaretto
    2   tsp. grated orange peel
    2   c. all-purpose flour
    ½   tsp. baking powder
    1   c. chopped dried huckleberries (recipe page 238)
    1   c. chopped almonds

Directions:

1. Preheat oven to 325 degrees F.
2. In large bowl with mixer on medium speed, beat butter and sugar until smooth.
3. Add egg yolk, liqueur, and orange peel; beat until well blended.
4. In another bowl mix flour and baking powder; add to butter mixture, stir to mix, and then beat until well blended.
5. Gather dough into ball, divide in half, and flatten each portion into a disk; wrap each disk tightly in plastic wrap, and freeze until firm enough to roll without sticking, about 30 minutes.
6. Place dough on lightly floured surface; roll one disk at a time to about ¼ inch thick.
7. With floured, 2-inch round cutter, cut out cookies; place about 2 inches apart on buttered cookie sheets.
8. Make 1½-inch-wide depression with finger in center of each cookie; continue with dough until all is used.
9. In small bowl beat egg white with 1 teaspoon water to blend; brush cookies with mixture and sprinkle or

arrange about ½ teaspoon chopped huckleberries and almonds onto top of each.

10. Bake for 15 minutes.
11. Remove from oven, and allow cookies to cool on sheets for 5 minutes; use wide spatula to transfer to wire racks to cool completely.

Yields: 3 dozen cookies.

Huckleberry Bars

These delicious bars are not only easy to make, but they make good lunch treats, too. They also freeze well.

Ingredients:

¾   c. butter, softened
1   c. brown sugar, firmly packed
1½ c. all-purpose flour
½   tsp. salt
½   tsp. baking soda
1½ c. rolled oats
9   oz. huckleberry preserves (recipe page 197)

Directions:

1. Preheat oven to 400 degrees F.
2. In medium bowl cream together butter and sugar.
3. Add dry ingredients, and stir with fork to consistency of crumbs.
4. Pat half of crumb mixture into buttered 13 x 9 x 2-inch pan.
5. Spread with preserves and sprinkle with remaining crumb mixture.
6. Bake for 25 minutes.
7. Remove from oven and cool on wire rack; cut into squares before serving.

Yields: 12 to 14 servings.

Chocolate Drizzle Huckleberry-Filled Cookies

Huckleberry preserves fill these delicious pecan-coated cookies, and a chocolate drizzle dresses them up.

Ingredients:

- ½ c. butter, softened
- ½ c. sugar
- 1 lg. egg, separated
- 1 tsp. vanilla extract
- 1 c. all-purpose flour
- ½ tsp. salt
- 2 c. finely chopped pecans, divided
- ½ c. huckleberry preserves (recipe page 197)
- 1 c. semisweet chocolate morsels, melted

Directions:

1. In large mixing bowl beat butter with sugar until fluffy; add egg yolk and vanilla, beating until creamy.
2. Combine flour and salt; add to butter mixture, beating well.
3. Stir in 1 cup chopped pecans; cover and chill dough in refrigerator at least 30 minutes.
4. Preheat oven to 350 degrees F.
5. Shape dough into 1-inch balls; dip each ball in beaten egg white, roll in remaining chopped pecans, and place on greased cookie sheets 1 inch apart.
6. Press thumb gently into center of each ball, leaving an indention; fill with huckleberry preserves.
7. Bake for 17 to 18 minutes or until lightly browned.
8. Remove from oven, and cool 1 minute on cookie sheets; remove to wire racks and cool completely.
9. Drizzle melted chocolate over cooled cookies, and let sit to harden chocolate.

Yields: 3 dozen cookies.

Huckleberry Caramel Nut Squares

If you enjoy the rich taste of caramel and nuts, you will enjoy these bars. The huckleberries and nuts blend in very well with the caramel flavor.

Ingredients for crust:

- ½ c. butter, softened
- ½ c. brown sugar, firmly packed
- 2 egg yolks, beaten
- 1¼ c. all-purpose flour

Ingredients for topping:

- 2 egg whites
- 1 c. brown sugar, lightly packed
- 1 tsp. vanilla extract
- 1 c. walnuts or pecans, chopped

Directions for crust:

1. In medium bowl cream together butter and brown sugar.
2. Add egg yolks and beat well.
3. Add flour and combine thoroughly.
4. Spread mixture in bottom of greased 8-inch square pan.

Directions for topping:

1. Preheat oven to 350 degrees F.
2. In medium bowl beat egg whites until stiff.
3. Blend in brown sugar and vanilla; fold in nuts.
4. Spread over crust and bake for 30 minutes.
5. Cool on wire rack before slicing to serve.

Yields: 16 servings.

Frosty Huckleberry Squares

This is a refreshing huckleberry dessert that can be made ahead and taken out in a hurry for unexpected company.

Ingredients for crust:

- 1 c. all-purpose flour
- ¾ c. hazelnuts, chopped
- ½ c. butter, melted
- ¼ c. brown sugar, packed

Ingredients for filling:

- 2 c. whipping cream, divided
- 2½ c. huckleberries, fresh or frozen
- 1 c. sugar
- 2 Tbs. lemon juice

Directions for crust:

1. Preheat oven to 350 degrees F.
2. In medium bowl combine flour, nuts, butter, and brown sugar to make crumb mixture.
3. Spread evenly into greased shallow baking pan.
4. Bake for 10 minutes, stir, and continue baking for another 10 minutes.
5. Sprinkle ⅔ of baked crumbs into 13 x 9 x 2-inch baking pan; reserve remaining crumbs for topping.

Directions for filling:

1. In large mixing bowl combine 1 cup whipping cream, huckleberries, sugar, and lemon juice.
2. Beat with electric mixer on high speed about 10 minutes or until stiff peaks form.
3. In small mixing bowl beat remaining 1 cup whipping cream into stiff peaks.
4. Fold into huckleberry mixture and spoon evenly over crumbs in baking pan.

5. Sprinkle remaining crumbs on top.
6. Cover and freeze until firm and ready to serve.

Yields: 12 to 14 servings.

Huckleberry Pecan Cookies

These are spicy bar cookies with a nice crunch of pecans plus the sweetness of huckleberry jam.

Ingredients:

1   c. butter, softened
1   c. sugar
2   egg yolks
1   tsp. vanilla extract
⅛   tsp. ground cardamom
¼   tsp. ground allspice
2   c. all-purpose flour
1   c. chopped pecans
½   c. huckleberry jam (recipe page 192)

Directions:

1. Preheat oven to 325 degrees F.
2. In large bowl cream butter until soft and fluffy.
3. Add sugar gradually, beating until light and fluffy; beat in egg yolks.
4. Sift cardamom, allspice, and flour together; gradually add to butter mixture and stir to combine well.
5. Stir in chopped pecans.
6. Spoon half of dough into greased 8-inch square pan, spreading evenly; top with huckleberry jam and cover with remaining dough.
7. Bake for 1 hour or until lightly browned.
8. Remove from oven and cool on wire rack in pan; cut into squares.

Yields: 2 dozen cookies.

Huckleberry Cheesecake Cookie Bars

My family loves cheesecake and cookies, and combined together they make a delicious treat.

Ingredients:

- ¼ c. butter, softened
- ¼ c. brown sugar, firmly packed
- ¼ c. all-purpose flour
- ¾ c. ground walnuts
- 8 oz. cream cheese, softened
- ¼ c. sugar
- 1 egg, lightly beaten
- 2 Tbs. lemon juice
- ½ tsp. pure vanilla extract
  huckleberry jam (recipe page 192)

Directions:

1. Preheat oven to 350 degrees F.
2. Cream butter and brown sugar; combine with flour and walnuts until mixture is crumbly.
3. Press into bottom of buttered 9 x 9-inch baking pan.
4. Bake for about 12 minutes; cool slightly.
5. Beat cream cheese, sugar, egg, lemon juice, and vanilla.
6. Spread evenly over crust and top with huckleberry jam.
7. Return to oven and bake another 25 minutes.
8. Cool to room temperature and cut into 2-inch squares.

Yields: 20 cookie bars.

Did You Know? . . . .

Did you know that huckleberry plants in Montana can be anywhere from 2 inches tall with berries the size of match heads to shrubs up to 6 feet tall with pea-size and larger berries?

Huckleberry Cranberry Nut Bars

These bars are a great take-along snack packed with the goodness of grains and fruit.

Ingredients:

- 1 c. all-purpose flour
- 1 c. quick-cooking oats
- ⅔ c. brown sugar, firmly packed
- 2 tsp. baking soda
- ½ tsp. salt
- ½ tsp. ground cinnamon
- ⅔ c. buttermilk
- 3 Tbs. vegetable oil
- 2 lg. egg whites
- ⅔ c. dried huckleberries (recipe page 238)
- ½ c. dried cranberries
- ¼ c. chopped nuts
- 2 c. flaked coconut

Directions:

1. Preheat oven to 375 degrees F.
2. Lightly grease 9-inch square baking pan.
3. In large mixing bowl combine flour, oats, brown sugar, baking soda, salt, and cinnamon; stir to blend.
4. Add buttermilk, oil, and egg whites; beat just until blended.
5. Stir in dried fruits and nuts.
6. Sprinkle half of coconut over bottom of prepared pan.
7. Spread batter evenly over top of coconut; sprinkle remaining half of coconut over top of batter.
8. Bake for 20 to 25 minutes or until cake tester inserted in center comes out clean.
9. Remove from oven, cool, and cut into bars.

Yields: 20 servings.

Huckleberry Drops

These make a great-tasting drop cookie that can be underbaked for a softer cookie, or leave them in the oven longer to enjoy a crisper one.

Ingredients:

3	c. all-purpose flour
½	tsp. salt
1	tsp. baking powder
½	tsp. baking soda
½	c. butter, softened
1	c. sugar
¾	c. brown sugar, firmly packed
1	egg, lightly beaten
2	Tbs. orange juice concentrate, thawed
¼	c. milk
1	c. walnuts, chopped or ground
2½	c. frozen huckleberries, thawed, rinsed

Directions:

1. Preheat oven to 375 degrees F.
2. Sift together flour, salt, baking powder, and baking soda; set aside.
3. Cream butter, sugars, and egg until light in consistency.
4. Mix orange juice concentrate with milk; add to creamed mixture.
5. Pat huckleberries dry on paper towels, and mix into dry ingredients along with nuts.
6. Stir dry ingredients with creamed mixture just until mixed.
7. Drop dough by teaspoonfuls onto greased baking sheet.
8. Bake for 10 to 12 minutes.
9. Remove from oven and transfer to wire racks to cool.

Huckleberry Macaroon Drops

These are quite delicious and a variation on traditional macaroons with the addition of the huckleberries.

Ingredients:

- 1 c. sugar
- ½ lb. almond paste
- 2 Tbs. condensed cream
- 1 c. powdered sugar
- 1 Tbs. cake flour
- 2 c. flaked coconut
- 2 Tbs. huckleberry leather, chopped fine (recipe page 240)
  vegetable oil
  powdered sugar

Directions:

1. In large mixing bowl mix sugar and almond paste until well blended; blend in cream.
2. Combine flour and powdered sugar; blend into almond paste mixture.
3. Add 2 drops of oil to chopped huckleberry leather to separate pieces, then add to mixture along with coconut flakes; mix well.
4. Drop by teaspoonfuls onto cookie sheets covered with parchment or wax paper.
5. Cover and let stand for 30 minutes or until firm; if desired, dust or roll in additional powdered sugar.

Yields: 2 dozen pieces.

Did You Know? . . . .

Did you know that huckleberries should be frozen in a tightly sealed container so their fragrance does not permeate everything in your freezer?

Huckleberry Pinwheels

These make attractive-colored pinwheels that are great tasting. With the addition of the grated orange peel, it makes for an extra-tasty treat.

Ingredients:

- 1   c. huckleberry jelly (recipe page 188)
- 1   Tbs. cornstarch
- ½   c. butter, softened
- ¾   c. sugar
- 1   egg
- 1¾ c. all-purpose flour
- 1   tsp. baking powder
- ¼   tsp. ground allspice
- 1   tsp. finely grated orange peel

Directions:

1. Combine huckleberry jelly and cornstarch in small saucepan; bring to boil over moderate heat, stirring constantly.
2. Remove from heat and refrigerate until cooled.
3. Cream butter and sugar; add egg and beat until fluffy.
4. Stir in flour, baking powder, and allspice, blending well; add grated orange peel.
5. Cover with plastic wrap, and chill in refrigerator for at least 1 hour.
6. On lightly floured surface, roll out chilled dough into 16 x 8-inch rectangle.
7. Spread cooled huckleberry filling on dough to within ½ inch of edges.
8. Starting with long edge, roll up.
9. Cut in half, wrap each piece in plastic, and refrigerate for 2 hours.
10. Preheat oven to 375 degrees F.
11. With sharp knife cut cookies about ½ inch thick, and place on greased cookie sheet 2 inches apart.

12. Bake for 10 to 12 minutes.
13. Remove from oven and gently lift with spatula onto wire racks to cool.

Yields: 30 cookies.

Huckleberry Thumbprint Cookies

When I was a child, my mom made thumbprint cookies which I really enjoyed. This is a great-tasting version using your favorite huckleberry jam.

Ingredients:

- 1   c. butter, softened
- 1   c. brown sugar, firmly packed
- 2   eggs, separated
- 2   c. all-purpose flour
- 1   c. ground walnuts
- 1   c. huckleberry jam (recipe page 192)

Directions:

1. In large bowl cream butter and sugar; add egg yolks and mix until smooth.
2. Combine with flour and refrigerate until firm.
3. Preheat oven to 325 degrees F.
4. Remove dough from refrigerator and form into 1-inch balls.
5. Lightly beat egg whites with fork until thin.
6. Dip dough balls into egg whites, then roll in ground walnuts.
7. Place on cookie sheet and bake just for 5 minutes.
8. Remove from oven and make thumbprint in center of each cookie, then fill with small amount of huckleberry jam.
9. Return to oven and bake 10 minutes longer.
10. Allow to cool before removing from cookie sheet.

Yields: 4½ dozen cookies.

Huckleberry Shortbread Bars

These are very tasty huckleberry bars that can be made year round with your favorite huckleberry jam.

Ingredients for bottom layer:

  1½ c. all-purpose flour
  ½   c. sugar
  ½   c. butter
  ¾   c. huckleberry jam (recipe page 192)

Ingredients for top layer:

  2   Tbs. all-purpose flour
  ¼   tsp. salt
  ¼   tsp. baking soda
  2   eggs, lightly beaten
  ½   c. brown sugar, firmly packed
  1   tsp. pure vanilla extract
  1¼ c. chopped walnuts

Directions for bottom layer:

1. Preheat oven to 350 degrees F.
2. In medium bowl combine flour with sugar; cut in butter until mixture becomes finely crumbled.
3. Press into bottom of greased 9 x 9-inch baking pan.
4. Bake for 20 minutes; remove from oven and let cool in pan. (Leave oven on.)
5. Spread huckleberry jam over crust.

Directions for top layer:

1. Mix flour, salt, and baking soda.
2. Combine eggs, brown sugar, and vanilla; stir in flour mixture.
3. Fold in walnuts and spoon over jam layer.
4. Return to oven and bake 20 minutes longer.

5. Cool in pan, dust with powdered sugar if desired, and cut into bars.

Yields: 20 servings.

Huckleberry Walnut Bars

These bars make a great-tasting after-school snack which keeps very well.

Ingredients:

1   c. sugar
1   c. butter, room temperature
1   egg
½   tsp. almond extract
2¼ c. all-purpose flour
1   c. chopped walnuts
    huckleberry topping (recipe page 250)

Directions:

1. In large bowl cream sugar and butter; add egg and almond extract.
2. Combine flour and walnuts; add to creamed mixture and blend well.
3. Reserve ⅓ of mixture; press remaining ⅔ onto bottom of greased 9 x 9-inch baking pan; refrigerate for at least 1 hour.
4. Preheat oven to 350 degrees F.
5. Remove from refrigerator, and spread huckleberry topping over cold dough in pan.
6. Sprinkle reserved crumb mixture over top, and bake for about 45 minutes.
7. Remove from oven, and cool completely before cutting into bars or squares.

Yields: 20 servings.

Huckleberry Oatmeal Bars

Oatmeal and huckleberries make a great taste combination. The coconut adds texture and flavor.

Ingredients:

   1¼ c. all-purpose flour
   1½ c. quick-cooking oats
   ½   c. sugar
   ½   tsp. baking soda
   ⅓   c. butter, melted
   2   tsp. pure vanilla extract
   1   c. coconut, shredded
        huckleberry jam (recipe page 192)

Directions:

1. Preheat oven to 350 degrees F.
2. Combine flour, oats, sugar, soda, butter, and vanilla to make crumb mixture.
3. Reserve 1 cup of mixture; press remaining amount onto bottom of greased 13 x 9 x 2-inch baking pan.
4. Spread huckleberry jam onto this mixture; sprinkle coconut and reserved crumb mixture over top.
5. Bake for about 25 minutes.
6. Cool completely before cutting into bars.

Yields: 2 dozen bars.

Did You Know? . . . .

Did you know that the globe huckleberry has three color phases: dark red, blue-black or purple-black, and blue-black with a whitish coating like the one on plums?

Huckleberry Delights Cookbook

A Collection of Huckleberry Recipes
Cookbook Delights Series Book 6

Desserts

Table of Contents

Apple Huckleberry Crisp

Our family loves all kinds of crisps. This one is a terrific combination of apples and fresh huckleberries. It is a great dessert that is best served warm with ice cream.

Ingredients for filling:

- 1   tsp. canola oil
- 4   Granny Smith apples, peeled, cored, sliced
- 1   Tbs. lemon juice
- 1   tsp. ground cinnamon
- ¼   c. sugar
- 1   c. frozen huckleberries

Ingredients for topping:

- 1   c. rolled oats
- ⅓   c. unbleached all-purpose flour
- ¼   c. brown sugar, firmly packed
- 4   tsp. canola oil
- ½   tsp. ground cinnamon
- 1   tsp. orange juice

Directions for filling:

1. Lightly oil 11 x 9 x 2-inch baking dish with canola oil.
2. Mix sliced apples with lemon juice, cinnamon, and sugar.
3. Press into baking dish and sprinkle frozen berries on top.

Directions for topping:

1. Preheat oven to 350 degrees F.
2. In medium bowl mix together oats, flour, brown sugar, oil, cinnamon, and orange juice.
3. Sprinkle over top of apples and berries in dish.
4. Bake for 30 minutes or until apples test done.
5. Remove from oven, and cool in pan on wire rack before cutting to serve.

Yields: 20 to 24 servings.

Huckleberry Baked Indian Pudding

I can just imagine the Native Americans picking huckleberries and making Indian pudding. Huckleberries and cornmeal go great together. Of course the Native Americans did not have fresh oranges and ginger at their fingertips, but they add a nice flavor.

Ingredients:

- 2 c. milk
- ¼ c. sugar
- ¼ c. stone-ground white cornmeal
- 1 lg. egg, lightly beaten
- 1 tsp. grated orange peel
- ½ tsp. ground ginger
- ¼ tsp. ground cinnamon
- ¼ c. light molasses
- ¼ c. brown sugar, firmly packed
- 1 c. huckleberries, fresh or frozen

Directions:

1. Preheat oven to 300 degrees F.
2. In heavy saucepan mix milk with sugar.
3. Place over medium-high heat and stir until milk is simmering; gradually sprinkle in cornmeal and whisk until smooth.
4. In small bowl whisk together egg, orange peel, ginger, cinnamon, molasses, and brown sugar.
5. Whisk in small amount of cornmeal mixture.
6. Return whole mixture to saucepan of cornmeal; stir to blend.
7. Pour mixture over berries placed in bottom of greased 1½-quart baking dish.
8. Bake for 45 to 55 minutes or until knife inserted into center of pudding comes out clean.
9. Remove from oven, and place on wire rack to cool to room temperature.

Yields: 8 to 10 servings.

Apple Huckleberry Tart

This tart makes an attractive dessert. The combination of a homemade tart shell, huckleberries, and apples is delicious. It is great served warm with homemade ice cream.

Ingredients for tart dough:

2	c. all-purpose flour
¼	tsp. salt
1	tsp. sugar
6	oz. unsalted butter, cut into pieces
7	Tbs. water

Ingredients for filling:

6	c. firm, tart baking apples (like Granny Smith), peeled, sliced
1	c. huckleberries
⅓	c. sugar
2	Tbs. melted butter

Directions for tart dough:

1. In medium bowl mix flour, salt, and sugar.
2. Cut in 2 ounces of butter, and carefully work into flour mixture to crumbly consistency.
3. Cut in remaining 4 ounces of butter and mix well, about 5 minutes, leaving butter in bigger pieces.
4. Add water 1 tablespoon at a time until dough just clings together.
5. Wrap in plastic and let rest about 15 minutes.
6. Roll out dough very thin, about ⅛ inch.
7. Place dough into 9-inch tart pan (edges of dough will overlap sides about 1½ inches); set aside.

Directions for filling:

1. Preheat oven to 375 degrees F.
2. In large bowl toss apples, huckleberries, and sugar together.
3. Spoon mixture onto dough; fold edges of dough over fruit.

4. Brush melted butter over edges.
5. Bake for 35 to 40 minutes, until dark and crispy.
6. Remove from oven and slide cooked tart onto rack to cool.
7. Slice and serve warm with ice cream.

Yields: 8 servings.

Almond Crème with Huckleberries

Nothing beats the combination of huckleberries and cream.
This makes an interesting presentation that is also very flavorful.

Ingredients:

¼   c. cold water
1½ env. unflavored gelatin
½   c. boiling water
1   can evaporated milk
½   c. sugar
½   tsp. vanilla extract
½   tsp. almond extract
3   c. fresh huckleberries
½   can sweetened condensed milk
     Sweetened Whipped Cream (recipe page 167)

Directions:

1. Sprinkle gelatin over cold water; let stand a few minutes.
2. Pour boiling water in medium mixing bowl; add gelatin mixture and stir until dissolved.
3. Combine evaporated milk, sugar, and extracts; stir into gelatin.
4. Divide into 6 individual serving bowls, and chill at least 3 hours.
5. Wash and pat dry huckleberries.
6. Divide berries over chilled crème, pour sweetened condensed milk on top of berries, and top with whipped cream.

Yields: 6 servings.

Deluxe Huckleberry Mousse

This mousse makes a delicious, light after-dinner dessert. Our family loves it and so will yours!

Ingredients:

¼   c. cold water
1   env. unflavored gelatin
3   c. frozen huckleberries, thawed with juice
1½ c. sugar
1   c. plain yogurt
1   c. heavy cream, whipped
2   Tbs. amaretto liqueur
     mint leaves for garnish

Directions:

1. Sprinkle gelatin over cold water and let stand.
2. In saucepan mix together huckleberries and sugar; cook mixture over moderate heat about 5 minutes, stirring constantly.
3. Add softened gelatin and stir to dissolve.
4. Cool to room temperature and purée in blender.
5. Fold in yogurt and whipped cream, then amaretto.
6. Spoon into 8 wine glasses and chill 1 hour.
7. Just before serving garnish with mint leaf.

Yields: 8 servings.

Did You Know? . . . .

Did you know that you only need to purchase a permit to pick huckleberries in the Flathead National Forest in Montana if you pick more than 10 gallons per person? If you sell the berries you pick or pick more than the 10-gallon limit, you will be considered a commercial picker and must purchase a permit.

Rice Pudding with Huckleberries

Rice pudding is known as a comfort food. This is a delightful variation on an old favorite.

Ingredients:

¾ c. short-grain rice
4   c. milk
½ c. sugar
1   dash nutmeg
1   c. fresh huckleberries
½ c. heavy cream
1   tsp. vanilla extract

Directions:

1. Combine rice and milk in saucepan or in top of double boiler, and cook until very soft, about 45 minutes.
2. Stir in sugar and nutmeg; cool slightly.
3. Fold in huckleberries; pour into broiler-proof dish.
4. Whip cream and vanilla together; spread over top of pudding in dish.
5. Place under broiler and brown top slightly.
6. Sprinkle a few additional berries on top for garnish.
7. Serve at once.

Yields: 8 servings.

Did You Know? . . . .

Did you know that the best way to freeze huckleberries is to spread them in a single layer on a cookie sheet lined with paper towels? When they are frozen, then you can transfer the berries to a container. They will not be frozen in one big clump this way, so you will be able to dip them out in whatever quantity needed.

Fresh Huckleberry Custard Parfait

Custard and berries go very well together. This is a very flavorful and attractive dessert.

Ingredients:

¼   c. plus 1 Tbs. sugar, divided
2   lg. eggs, slightly beaten
1½ c. milk
1   tsp. vanilla extract
1   tsp. grated orange peel
½   tsp. grated lemon peel
¼   tsp. ground nutmeg
⅓   c. whipping cream
2   c. fresh huckleberries
    Sweetened Whipped Cream (recipe page 167)

Directions:

1. In top pan of double boiler, combine ¼ cup sugar with eggs and milk.
2. Over medium heat stir over hot water, not boiling, until custard coats wooden spoon.
3. Remove from heat; cool to warm.
4. Stir in vanilla, orange and lemon peel, and nutmeg.
5. Whip cream with 1 tablespoon sugar until cream holds its shape; fold into custard mixture.
6. Arrange alternate layers of huckleberries and custard in parfait glasses; chill.
7. When ready to serve, place dollop of sweetened whipped cream on top of each serving.

Yields: 4 servings.

Did You Know? . . . .

Did you know that the huckleberry is sometimes called "the poor man's blueberry"?

Huckleberries Romanoff

If you enjoy fresh berries, you will enjoy this refreshing way to use huckleberries.

Ingredients:

- ½ c. orange liqueur
- 1 c. sour cream
- ½ pt. vanilla bean ice cream
- 2 c. sweetened whipped cream (recipe page 167)
- 2 c. huckleberries
- 2 c. raspberries
- 2 c. blackberries
- 2 c. loganberries
- 1 pt. strawberries, hulled, halved
- 2 oz. chocolate curls
- 2 Tbs. chiffonade of mint
  powdered sugar in shaker

Directions:

1. In large mixing bowl whisk orange liqueur and sour cream together.
2. Stir in ice cream until blended; fold in whipped cream.
3. Wash berries and drain on paper towels; arrange on platter.
4. Spoon cream mixture over top of berries.
5. Garnish with chocolate curls, mint, and powdered sugar.

Yields: 10 to 12 servings.

Did You Know? . . . .

Did you know that with many recipes you do not need to thaw frozen huckleberries but simply add them frozen? You may need to extend the cooking or baking time for some recipes, though.

Huckleberry Coffee Ice Cream

Our family enjoys homemade ice cream. This is an interesting combination for those who love coffee.

Ingredients:

4 c. huckleberries
1½ c. sugar
1 env. unflavored gelatin
1 Tbs. instant coffee granules
4 c. light cream, divided
1 egg, slightly beaten
1 tsp. vanilla extract

Directions:

1. Place huckleberries and sugar in 2-quart saucepan.
2. Bring to boil over moderate heat, stirring constantly, then cook about 2 minutes longer.
3. Remove from heat and transfer to bowl; cool to room temperature then chill.
4. Sprinkle gelatin and instant coffee over half the cream.
5. Stir over low heat only until gelatin and coffee dissolve.
6. Stir small amount of this hot mixture into egg, then return egg mixture to hot liquid; cook, stirring until slightly thickened, about 1 minute.
7. Cool to room temperature, then chill.
8. Combine with remaining light cream, vanilla, and chilled huckleberry mixture.
9. Freeze in ice cream freezer according to manufacturer's directions.

Yields: About 2½ quarts.

Did You Know? . . . .

Did you know that Huckleberry Hound first appeared in 1958?

Huckleberry Frozen Parfait

This is a creamy, fruity frozen dessert that is nice on a hot summer day.

Ingredients:

1½ c. huckleberries
¼ c. sugar
2 Tbs. cornstarch
¼ c. water
3 Tbs. lemon juice
8 oz. plain yogurt
3 peaches, coarsely chopped
fresh mint leaves, chopped, for garnish

Directions:

1. In saucepan stir together huckleberries and sugar.
2. Mix cornstarch and water until smooth; add to huckleberry mixture.
3. Cook over medium heat, stirring constantly, until mixture comes to a boil; boil 1 minute.
4. Remove from heat; stir in lemon juice, then cool.
5. Fold yogurt into cooled huckleberries.
6. In parfait glasses, alternately layer huckleberry mixture with peaches; freeze.
7. Remove from freezer ½ hour before serving.
8. When ready to serve, sprinkle with chopped mint leaves for garnish.

Yields: 4 to 6 servings.

Did You Know? . . . .

Did you know that during World War II, British pilots ate bilberries (huckleberries) before night flights to enhance their vision?

Tasty Huckleberry Ice

This huckleberry ice is light, refreshing, and colorful. Enjoy.

Ingredients:

- 2 c. frozen huckleberries
- 4 Tbs. fresh lemon juice
- ½ c. sugar
- ½ c. water
- 4 oranges (optional serving containers)
  mint leaves for garnish
  lemon slices for garnish

Directions:

1. Process huckleberries, lemon juice, sugar, and water in food processor or blender until puréed.
2. Serve immediately, or cover and place in freezer.
3. To serve, thaw for 20 to 30 minutes and stir to desired consistency.
4. Serve in hollowed out oranges or small dessert bowls.
5. Garnish with mint leaves or thin lemon slices.

Yields: 4 servings.

Wild Huckleberries À La Crème Brûlée

Crème Brûlée is a popular dessert, and this one is made extra special with the addition of wild huckleberries.

Ingredients:

- 2 c. huckleberries
- ⅔ c. sour cream
- ½ c. vanilla yogurt
- ⅛ tsp. ground cinnamon (optional)
- ⅓ c. brown sugar, firmly packed

Directions:

1. Divide huckleberries among 4 heatproof ramekins.
2. Combine sour cream, yogurt, and cinnamon; spread over huckleberries, covering completely.
3. Sprinkle brown sugar over cream mixture, and broil 3 inches from heat source until sugar bubbles and caramelizes, approximately 3 to 5 minutes.
4. Serve while topping is still hot.

Yields: 4 servings.

Huckleberry Banana Dessert

This is a simple fruit dessert that is delicious.

Ingredients:

2   c. fresh huckleberries
2   med. bananas, sliced
3   Tbs. powdered sugar
1   dash ground nutmeg
2   c. whipped cream

Directions:

1. Wash and dry huckleberries; divide among 6 to 8 serving bowls.
2. Mix sliced bananas with sugar and nutmeg; fold in whipped cream until mixed in.
3. Spoon over berries; chill in refrigerator for at least 1 hour or until ready to serve.
4. When ready to serve sprinkle additional nutmeg over top of each serving, if desired.

Yields: 6 to 8 servings.

Huckleberry Meringue Baskets

My children especially enjoy baked meringue. These baskets are worth the extra effort. If you do not want to take the trouble to make the basket shapes, you can simply spoon into three 8-inch cake pans.

Ingredients:

- 3 egg whites
- 1 c. sugar
- 8 oz. cream cheese, softened
- ⅔ c. powdered sugar
- 3 Tbs. huckleberry purée
- 1 c. Sweetened Whipped Cream (recipe page 167) fresh huckleberries for garnish

Directions:

1. Preheat oven to 160 degrees F.
2. Place egg whites and sugar into heatproof bowl.
3. Set bowl over pan of simmering water.
4. Gently whisk until sugar is completely dissolved and egg whites are hot; about 5 to 6 minutes.
5. Remove bowl from simmering water and beat egg whites until stiff peaks form.
6. When cooled to room temperature, transfer into pastry bag fitted with ½-inch star tip.
7. Line baking sheet with wax paper; pipe meringues in coiled circles about 3 inches across.
8. Pipe out another ring along outside edge for top. (You should end up with 10 baskets.)
9. Place baking sheet in preheated oven. (The low temperature is very important. Meringue baskets will stay in oven 4 hours to dry out thoroughly, yet they should remain white.)
10. When done, remove from oven and cool to room temperature.
11. Prepare filling by mixing cream cheese, powdered sugar, and huckleberry purée.

12. Fold in the whipped cream.
13. Pipe filling into baskets and garnish with fresh huckleberries; serve immediately.

Yields: 10 baskets.

Huckleberry Bread Pudding

My mom always made us bread pudding, which my children still enjoy. This is a tasty version of this comfort food.

Ingredients:

12 slices white bread
2   c. huckleberries, rinsed, drained
3   eggs, beaten
1½ c. warm milk
1   can sweetened condensed milk
⅓ c. butter, melted
1   tsp. vanilla extract
½ tsp. almond extract
    powdered sugar

Directions:

1. Preheat oven to 350 degrees F.
2. Cut bread slices into cubes; combine with huckleberries.
3. Place in greased 9 x 9 x 2-inch baking pan.
4. In medium bowl mix together eggs, milk, condensed milk, butter, and extracts; blend well.
5. Pour over bread and huckleberry mixture.
6. Bake for 50 minutes.
7. Remove from oven and cool on wire rack.
8. Sprinkle with powdered sugar just before serving.

Yields: 8 to 10 servings.

Huckleberry Panna Cotta

The huckleberries and brandy make an interesting combination.

Ingredients:

- 1   c. huckleberries
- 1   c. milk, divided
- 1½ Tbs. unflavored gelatin
- 4½ c. cream
- ½   c. sugar
- 1   tsp. brandy

Directions:

1. Purée huckleberries with ½ cup milk.
2. Add gelatin to berry mixture to soften.
3. In saucepan heat cream, remaining milk, and sugar to a boil, stirring constantly.
4. When cream reaches a boil, add huckleberry mixture and brandy; heat through to activate gelatin.
5. Ladle into molds; refrigerate 2 to 3 hours to set.

Yields: 8 to 10 servings.

Huckleberry Raspberry Swirl Sherbet

This sherbet is a very colorful, delicious, light, and fruity finish to any meal.

Ingredients:

- 1   c. huckleberries, thawed, with juice
- ½   c. sugar
- 1   Tbs. lemon juice
- ½   gal. raspberry sherbet
       few drops of almond extract

Directions:

1. Combine berries, sugar, lemon juice, and almond extract in saucepan.
2. Bring to a boil over moderate heat, stirring constantly; cook until berries are soft.
3. Cool, then purée in blender until smooth; pour into bowl, then chill.
4. Allow raspberry sherbet to soften in refrigerator about 20 minutes.
5. Place in 1-gallon plastic container; pour chilled huckleberry mixture over sherbet, and gently fold in to create marble effect.
6. Cover and freeze until firm before serving.

Yields: 8 to 10 servings.

Sweetened Whipped Cream

This is a delicious topping on many desserts. Enjoy.

Ingredients:

1   c. heavy cream
¼   c. sugar
1   tsp. vanilla extract

Directions:

1. Whip cream until almost stiff.
2. Add sugar and vanilla; beat until cream holds peaks.

Did You Know? . . . .

Did you know that the taste of huckleberries ranges from tart to sweet, with a flavor comparable to that of blueberries, especially in the varieties that are blue/purple colored?

Amaretto Huckleberry Dessert

This is a terrific combination of flavors, making this a much sought after, delicious dessert.

Ingredients:

- 2   c. frozen huckleberries
- 1   c. sugar
- 1   Tbs. cornstarch
- 2   Tbs. amaretto liqueur
  coffee-flavored ice cream
  Sweetened Whipped Cream (recipe page 167)

Directions:

1. Combine huckleberries, sugar, and cornstarch in saucepan.
2. Bring to a boil over moderate heat, stirring continuously; boil 2 to 3 minutes.
3. Remove from heat and cool to room temperature.
4. Stir in amaretto liqueur.
5. Place 2 scoops of coffee ice cream into serving bowls; pour huckleberry mixture over top, and garnish with whipped cream.

Yields: 8 servings.

Did You Know? . . . .

Did you know that naturopathic practitioners use the huckleberry leaf to treat sugar diabetes and disorders of the kidneys and gallbladder?

Did you know that huckleberries contain tannins and anthocyanins which improve vision and stimulate circulation by strengthening capillaries?

Huckleberry Delights Cookbook

A Collection of Huckleberry Recipes
Cookbook Delights Series Book 6

Dressings, Sauces, and Condiments

Table of Contents

Easy Huckleberry Barbeque Sauce

If you do not have time for the more complicated version of barbeque sauce, try this simple, easy-to-make recipe. This sauce is excellent with salmon.

Ingredients:

    1   c. frozen huckleberries, thawed, with juice
    4   Tbs. honey
    18 oz. barbeque sauce

Direction:

1. Combine huckleberries and honey in saucepan.
2. Bring to a boil over moderate heat, stirring constantly.
3. Reduce heat and cook 2 minutes longer.
4. Cool, then purée in blender.
5. Mix with any barbeque sauce; cover and refrigerate until ready to use.

Huckleberries in Spiced Maple Syrup

For those of you who enjoy spices, this recipe adds a fragrant touch to fresh huckleberries.

Ingredients:

    ½   c. water
    ¼   c. maple syrup
    ½   c. brown sugar, firmly packed
    3   strips orange peel (6-in. length), pared
    1   stick cinnamon
    1   Tbs. vanilla extract
    4   c. fresh huckleberries

Directions:

1. Combine water, maple syrup, brown sugar, orange peel, and cinnamon in saucepan.
2. Cook, covered, over low heat until sugar is completely dissolved, stirring constantly.
3. Add vanilla and huckleberries.
4. Simmer slowly until berries are soft but not mushy, stirring carefully.
5. Remove from heat and cool.
6. Remove orange peel and cinnamon stick.
7. Use as syrup over vanilla- or coffee-flavored ice cream.

Fresh Huckleberry Sauce

This is a thick and flavorful huckleberry sauce that can be used on desserts, meats, or most anything you want to try it on. It can also be served on its own as a dessert.

Ingredients:

½   c. sugar
1½ Tbs. cornstarch
2   c. huckleberries
⅓   c. water
2   Tbs. lemon juice
    Sweetened Whipped Cream (recipe page 167)

Directions:

1. In saucepan combine sugar and cornstarch; stir in berries.
2. Add water and lemon juice.
3. Stir while cooking over medium heat until thickened.
4. Spoon over your favorite food while warm.
5. Sprinkle with dash of nutmeg, and add a fresh pansy or sprig of peppermint herb on plate as garnish.
6. Or, to serve as dessert, cool in refrigerator, spoon into bowls, and top with sweetened whipped cream.

Huckleberry Barbeque Sauce

This makes a flavorful huckleberry barbeque sauce. With it you can add the great taste of huckleberry to your favorite barbequed foods!

Ingredients:

2   qt. huckleberries, fresh or frozen
1½ c. celery, chopped fine
1½ c. onion, chopped fine
1½ c. red pepper, chopped fine
1   clove garlic, minced well
1   carrot, shredded fine
½   c. honey or to taste
2   Tbs. molasses
1   c. vinegar
1½ tsp. salt
1½ tsp. pepper
1   Tbs. paprika
1   tsp. ground cinnamon
⅛   tsp. ground nutmeg or to taste
⅓   tsp. dry mustard or to taste
⅛   tsp. ground cloves or to taste
⅛   tsp. celery seed or as desired
⅛   tsp. ground ginger or to taste
    cayenne pepper to taste

Directions:

1. Purée huckleberries (thaw, if frozen, but do not drain).
2. Combine celery, onion, red pepper, garlic, and carrot in large saucepan; add honey, molasses, and vinegar, stirring well.
3. Add salt and pepper; blend in spices.
4. Add puréed berries and mix well.
5. Simmer over medium heat, stirring occasionally, until thickened.
6. Serve with roasted or grilled turkey, chicken, or pork.

Yields: 4 cups.

Huckleberry Vinaigrette

This salad dressing is great as it is, or try the creamy variation with your choice of mayonnaise, yogurt, or sour cream.

Ingredients:

- ¼   c. olive oil
- 3   Tbs. huckleberry balsamic vinegar (recipe page 248)
- ½   tsp. salt
- ⅛   tsp. ground black pepper

Directions:

1. Combine oil and vinegar; salt and pepper to taste.
2. Serve over your favorite foods.
3. Variation: For creamier dressing, stir in 1 tablespoon mayonnaise, plain yogurt, or sour cream.

Yields: ½ cup.

Huckleberry Sauce

This is a very simple sauce to make. Use it right away or freeze it for future use.

Ingredients:

- 1   pkg. huckleberries frozen in syrup (16 oz.)
- 2   Tbs. fresh lemon juice
- ½   c. sugar

Directions:

1. Thaw huckleberries and purée in blender.
2. Add fresh lemon juice and sugar to purée.
3. If not using immediately, sauce may be covered and refrigerated until ready to use, up to 2 days.

Yields: 2 cups.

Huckleberry Salsa

Serve this unique salsa as a condiment or topping with any foods desired.

Ingredients:

- 1 pt. huckleberries
- 1 pt. strawberries
- ¼ c. sugar
- 3 Tbs. minced sweet onion
- 1 Tbs. raspberry vinegar or lemon juice
- 1 tsp. freshly ground black pepper
- ¼ c. sliced or slivered almonds, toasted
  hot pepper sauce to taste

Directions:

1. Rinse huckleberries and strawberries; dry on paper towels.
2. Sort over berries; hull strawberries and cut into quarters.
3. In bowl combine all ingredients except almonds.
4. Mix well and refrigerate for at least 1 hour.
5. Just before serving, stir in almonds.

Yields: 3½ cups.

Huckleberry Apple Chutney

Try this different version of the usual chutney. Your family and guests will rave about this one, for sure!

Ingredients:

- 2 c. huckleberries
- 6 green apples, cored, peeled, cubed
- ½ c. water
- ½ c. cider vinegar
- 1 lg. onion, chopped
- 1 c. sugar
- 1 tsp. salt
- 1 tsp. ground cinnamon

1 tsp. ground ginger
1 pinch ground cloves

Directions:

1. **Combine** huckleberries, apples, water, vinegar, and onion in 4-quart cooking pot; stir well.
2. **Heat** completely then add sugar, salt, cinnamon, ginger, and cloves; cook over moderate heat about 1½ hours, adding more water if necessary.
3. **If** not serving immediately, cool, cover, and store in refrigerator for up to 3 days.
4. **When** ready to use, warm through and serve with turkey, roast game, or with rice and your favorite Indian food.

Huckleberry Glaze

This is an ideal glaze for smoked meat, duck, pork, any game, chicken, or game hen.

Ingredients:

1 Tbs. butter
2 Tbs. minced shallots
½ oz. brandy
2 oz. crème de cassis liqueur
1 tsp. huckleberry jelly (recipe page 188) or 1 tsp. sugar and ½ c. huckleberries
½ c. veal stock or beef broth
¼ c. whole milk
  salt and pepper if desired

Directions:

1. In large sauté pan melt butter over medium-high heat; add shallots and cook until light brown.
2. Add brandy, crème de cassis liqueur, and jelly or huckleberry mix; reduce until almost dry.
3. Add veal or beef broth, and heat until reduced by half.
4. Remove from heat and swirl in milk; season to taste with salt and pepper if desired.
5. To serve, slice and arrange cooked meat on warm plate or platter, pour hot glaze over meat, and serve immediately.

Huckleberry Ketchup

This makes very flavorful ketchup with a wonderful potpourri of spices.

Ingredients:

- 2 Tbs. vegetable oil
- 1 lg. clove garlic, crushed
- 1 Tbs. minced fresh ginger
- 1 med. onion, finely chopped
- 2 pt. huckleberries
- 1 c. fresh tomato, peeled, seeded, chopped
- 2 lg. purple plums, pitted, chopped
- ¼ c. dark brown sugar, firmly packed
- 1 Tbs. huckleberry vinegar (recipe page 179)
- 1 Tbs. fresh lemon juice
- 1 med. dried chili pepper, crumbled
- 1 tsp. ground cinnamon
- 1 tsp. ground cardamom
- 1 tsp. ground coriander
- 1 tsp. salt
- 1 tsp. freshly ground mixed peppercorns (white, green, red, and black)
- zest of 1 lemon, cut into julienne strips

Directions:

1. Heat oil in 2-quart or larger, heavy-bottom saucepan; add garlic and ginger, and cook over low heat for 2 minutes.
2. Add onion; cook until soft and transparent, stirring often.
3. Add huckleberries, tomato, plums, brown sugar, vinegar, lemon juice and zest, chili pepper, spices, salt, and pepper, stirring well.
4. Cook over medium heat until mixture begins to simmer.
5. Reduce heat and simmer gently for 30 minutes.
6. Remove from heat and let mixture cool slightly.
7. Purée in food processor or blender.
8. Return purée to pan and heat, bringing mixture to a simmer; cook until thick, about 1 hour.
9. Pour into 2 sterile pint jars or containers.

10. Cover and cool.
11. Store in refrigerator for up to 4 weeks or freeze.

Yields: 2 pints.

Huckleberry Caramel Sauce

This sauce is great served over your favorite ice cream or pound cake. It is also delicious on a side dish of sweet potatoes!

Ingredients:

    ½ c. sugar
    ¼ c. water
    2 Tbs. orange juice
    1 c. heavy or whipping cream
    ½ tsp. vanilla extract
    1 pt. fresh huckleberries
        fresh huckleberries for garnish

Directions:

1. Bring sugar and water to a boil in medium saucepan over medium-high heat.
2. Reduce heat to medium; cook, swirling pan occasionally, until mixture is caramel colored, 5 to 10 minutes.
3. Remove from heat, and carefully stir in orange juice with long-handled wooden spoon, stirring vigorously until blended. (Use caution as mixture will bubble.)
4. Return pan to heat; stir in heavy cream.
5. Bring to a boil; reduce heat, and simmer sauce until thickened and reduced to 1 cup, about 8 minutes.
6. Stir in vanilla; pour sauce into medium bowl and stir in huckleberries.
7. If not using immediately, cover and refrigerate for up to 4 days.

Yields: 8 servings.

Huckleberry Nutmeg Sauce

Nutmeg and sherry add a warm and delicious flavor to this huckleberry sauce.

Ingredients:

- 1 c. water
- 1 c. sugar
- 2 Tbs. cornstarch
- ½ tsp. nutmeg, freshly grated
- 2 c. frozen huckleberries, thawed, with juice
- 3 Tbs. sherry wine

Directions:

1. In medium saucepan combine water, sugar, cornstarch, and nutmeg.
2. Cook over moderate heat, stirring constantly, until thick and bubbly.
3. Remove from heat; stir in huckleberries and sherry.
4. Place on burner again and bring back to boiling.
5. Remove from heat, cool slightly, and serve on desired food.

Yields: 3½ cups.

Wild Huckleberry Chutney

Try this spiced chutney with your poultry or main dish for a wonderful taste variation.

Ingredients:

- ½ c. raspberry vinegar
- ½ c. sugar
- 1 med. onion, minced
- ¼ tsp. fresh ginger, minced
- ⅛ tsp. ground cinnamon
- 1 tsp. lemon rind, minced

1   pinch salt
1   pinch cayenne pepper
3   c. huckleberries, divided
¼   c. dried cranberries

Directions:

1. Combine vinegar, sugar, onion, ginger, cinnamon, lemon, salt, and cayenne in saucepan; bring to a boil and simmer 15 minutes.
2. Add 1 cup huckleberries and the cranberries; simmer 20 minutes, stirring frequently.
3. Add remaining 2 cups huckleberries and simmer another 10 minutes.

Yields: 1 cup.

Huckleberry Vinegar

This is very flavorful vinegar, so use it in place of the plainer varieties. It also makes an attractive gift when put in a decorative bottle.

Ingredients:

1   c. huckleberry juice (recipe page 242)
½   c. cider vinegar
¼   c. water
2   Tbs. sugar

Directions:

1. In shaker jar combine juice, vinegar, water, and sugar; shake well.
2. Pour into sterilized bottle and refrigerate; best kept refrigerated between uses.

Yields: About 1 pint.

Huckleberry Relish

This relish is great with your favorite hamburger, hot dog, or polish sausage. For best results keep everything coarsely chopped.

Ingredients:

- ¼ c. Vidalia onion, diced
- ¼ c. balsamic vinegar
- 2 Tbs. sugar
- 1 Tbs. fresh lemon juice
- 1½ c. huckleberries, washed
  salt and pepper to taste

Directions:

1. If using food processor, coarsely chop 1 small onion wedge and measure ¼ cup.
2. Return to processor and add remaining ingredients, or dice onion and berries by hand and mix with remaining ingredients.
3. Let sit at room temperature for 20 minutes or until meal is served.

Yields: 1½ cups.

Huckleberry Basil Vinegar

This is a flavorful blend of huckleberry and basil. It also makes a great gift when packaged in a decorative jar.

Ingredients:

- 3 c. fresh huckleberries, crushed
- ½ c. torn fresh basil leaves, firmly packed
- 4 c. white vinegar
  fresh basil leaves (optional)

Directions:

1. Combine huckleberries and basil in ½-gallon or larger, sterilized, wide-mouth jar; set aside.
2. Place vinegar in medium, nonreactive saucepan, and bring to a boil.
3. Pour hot vinegar over huckleberry mixture; cover jar, and let stand at room temperature for at least 2 weeks or up to 2 months.
4. For gifts, strain vinegar through several layers of cheesecloth into decorative jars; discard huckleberry pulp.
5. Add a few fresh basil leaves to each jar if desired; seal jars with cork or airtight lid.

Yields: 4 cups.

Huckleberry Dessert Sauce

In this recipe huckleberries are simply thickened to make a sauce for desserts, side dishes, and even some of your favorite meats.

Ingredients:

1   c. sugar
1   Tbs. cornstarch
2   c. frozen huckleberries
¼   tsp. almond extract

Directions:

1. In large saucepan combine sugar and cornstarch.
2. Add huckleberries, and cook over moderate heat, stirring constantly.
3. Bring to rolling boil; remove from heat and stir in almond extract.
4. Serve while warm, or cool, cover, and refrigerate for up to 4 days.
5. Use over your favorite desserts, side dishes, or meat entrées.

Tart Huckleberry Butter

Use this simple huckleberry butter on English muffins, pancakes, or crepes.

Ingredients:

    1   c. blended huckleberries (whip berries in food
        processor or blender until smooth)

Directions:

1. Heat blended huckleberries to a boil in saucepan.
2. Turn heat to low and simmer, stirring occasionally, until mixture reaches desired thickness.

Yields: ⅓ cup.

Huckleberry Grape Dressing

The addition of grape juice makes this a very tasty dressing. It will perk up even the plainest dish.

Ingredients:

    3   Tbs. honey
    1   c. grape juice
    1   c. fresh huckleberries
    2   Tbs. shallots, minced
    1   c. extra-virgin olive oil
        sea salt to taste
        freshly ground black pepper to taste

Directions:

1. Warm honey slightly and place in food processor with grape juice, huckleberries, and shallots.

2. Drizzle olive oil into mixture slowly while machine is blending. (Emulsification is not critical.)
3. Take 1 ounce of dressing per person and a handful of spinach leaves, and wilt lightly in pan.
4. We often add bits of cooked applewood smoked bacon, roasted bell peppers, toasted pine nuts, and dollops of fromage blanc around the outside.

Jalapeño Huckleberry Barbeque Sauce

The jalapeño and mustard add a great flavor to this barbeque sauce.

Ingredients:

1   tsp. olive oil
¼   c. chopped onion
1   jalapeño, seeded, chopped
1   pt. fresh huckleberries
2   Tbs. rice wine vinegar
1½ Tbs. dark brown sugar
1   Tbs. Dijon mustard
¼   c. water

Directions:

1. In small saucepan sauté onions and jalapeño in olive oil over medium-high heat until limp, 2 to 3 minutes.
2. Add huckleberries, vinegar, brown sugar, mustard, and water, blending well.
3. Cook at low boil for 15 minutes, stirring often.
4. Remove from heat and cool to lukewarm; purée sauce in blender or food processor until smooth.
5. May be kept covered in refrigerator for up to 1 week if not ready to use at once.

Huckleberry Honey-Mustard Salad Dressing

This is a flavorful dressing recipe that is great with salads or meats and side dishes.

Ingredients for huckleberry dressing:

¼   c. frozen huckleberries, thawed, with juice
3   Tbs. honey
2   c. homemade honey-mustard dressing (recipe follows)

Ingredients for honey-mustard dressing:

1½ c. real mayonnaise
¼   c. Dijon mustard
½   c. honey

Directions for huckleberry dressing:

1. Combine huckleberries and honey in saucepan.
2. Bring to a boil over moderate heat, stirring constantly.
3. Lower heat and simmer a few minutes more until berries are soft.
4. Cool then purée in blender.
5. Chill then mix with honey-mustard dressing.
6. Cover and chill until ready to use.
7. Serve with your favorite green salad or foods of choice.

Directions for honey-mustard dressing:

1. Stir together all ingredients in small bowl until smooth.
2. Cover and refrigerate for 1 hour before serving.

Yields: About 2½ cups huckleberry dressing; 2¼ cups honey-mustard dressing.

Did You Know? . . . .

Did you know that the red huckleberry grows mostly at low elevations, from sea level up to a maximum of about 5,970 feet altitude?

Huckleberry Delights Cookbook

A Collection of Huckleberry Recipes
Cookbook Delights Series Book 6

Jams, Jellies, and Syrups

Table of Contents

A Basic Guide for Canning Jams, Jellies, and Syrups

1. Wash jars in hot, soapy water inside and out with brush or soft cloth.

2. Run your finger around rim of each jar, discarding any with cracks or chips.

3. Rinse well in clean, clear, hot water, using tongs to avoid burns to hands or fingers.

4. Place upside down on clean cloth to drain well.

5. Place lids in boiling water for 2 minutes to sterilize and keep hot until placing on rim of jar.

6. Immediately prior to filling each jar, immerse in very hot water with tongs to heat jar (avoids breakage of jar with hot liquid).

7. Fill jar to within 1 inch of top of rim or to level recommended in recipe.

8. Wipe rim with clean damp cloth to remove any particles of food, and check again for any chips or cracks.

9. With tongs, place lid from hot bath directly onto rim of jar.

10. Using gloves, cloth, or holders, tighten lid firmly onto jar with ring or use single formed lid in place of ring to cover inner lid. Do not tighten down too hard as it may impede sealing.

11. Place on protected surface to cool, taking care to not disturb lid and ring. A slight indentation of lid will be apparent when sealed.

12. Leave overnight until thoroughly cooled.

13. When cooled, wipe jars with damp cloth and then label and date each.

14. Store upright on shelf in cool, dark place.

Did You Know? . . . .

Did you know that huckleberries and bilberries can be distinguished from blueberries by cutting them in half? Ripe huckleberries and bilberries are colored purple throughout, while blueberries have white or greenish flesh.

Huckleberry Bits in Crabapple Jelly

This jelly is so delicious. The bits of huckleberries in the jelly give it an unusual and delightful texture.

Ingredients:

    8   c. fresh crabapples
    1   cinnamon stick (3-in. length)
    3   c. sugar
    2½ c. fresh huckleberries
         water as needed

Directions:

1. Remove stems and blossom ends from crabapples, and cut into quarters; place in large stainless steel or nonreactive pot.
2. Add enough water to cover apples by 2 inches; bring to a boil over medium-high heat.
3. Reduce heat to medium, and simmer for 10 to 15 minutes; apples should soften and change color.
4. Remove from heat, cool, and strain apples and juice through 2 or 3 layers of cheesecloth, recovering at least 4 cups of juice.
5. Discard pulp and pour juice back into pan; add cinnamon stick, bring to a simmer, and cook for 10 minutes.
6. Remove cinnamon stick, and skim off any foam that comes to the top.
7. Stir in sugar until completely dissolved; continue cooking at low boil until temperature reaches 220 to 222 degrees F.
8. Add huckleberries and boil for 1 minute longer.
9. Pour into sterilized jars, and process following canning directions on page 186.

Did You Know? . . . .

Did you know that huckleberry plants grow best in damp, acidic soil?

Easy Huckleberry Jam

This is an easy-to-make jam for all your seasonal huckleberries.

Ingredients:

5½ c. crushed huckleberries
1   pkg. powdered fruit pectin (2 oz.)
7   c. sugar

Directions:

1. Place berries in 6- to 8-quart cooking pot; remove seeds from half of berries by sieving, then return to pot of crushed berries.
2. Stir pectin into prepared berries; bring fruit and pectin mixture to a full boil, stirring constantly.
3. Remove from heat; add sugar, stirring in completely to prevent burning and sticking to pan.
4. Return to heat and return to full rolling boil, stirring constantly; boil for 1 minute.
5. Remove from heat; skim foam from surface.
6. Pour into sterilized jars, and process following canning directions on page 186.

Yields: 4 pints.

Huckleberry Jelly

For those of you who prefer clear jelly over jam, try this recipe. It will add another great purple color for your Easter dinner.

Ingredients:

5½ c. huckleberry juice (directions follow)
2   pkg. fruit pectin
2   Tbs. lemon juice
8   c. sugar
½   tsp. almond extract

Directions:

1. To prepare huckleberry juice, crush 7 cups of huckleberries.
2. Add 1 cup of water, and cook mixture in saucepan over low to moderate heat until soft.
3. Cool to room temperature, then squeeze through double thickness of cheesecloth; discard residue.
4. Combine huckleberry juice, powdered pectin, and lemon juice in kettle.
5. Bring to rolling boil over medium-high heat, stirring constantly.
6. Add sugar all at once; continue stirring until mixture returns to a full boil.
7. Boil for 1 minute; remove from heat.
8. Stir in almond extract; skim off foam.
9. Pour into sterilized jars, and process following canning directions on page 186.

Easy Huckleberry Syrup

Nothing is better than hot pancakes, waffles, or French toast and homemade huckleberry syrup. Adjust the sweetness to your taste. I think it is best not too sweet.

Ingredients:

2½ c. frozen huckleberries, thawed, with juice
1   c. sugar
1   c. light corn syrup

Directions:

1. Place huckleberries (including juice) and sugar in blender, and process at high speed.
2. Pour into saucepan and bring to a boil over moderate heat.
3. Add corn syrup and cook a little while longer.
4. Pour into bottle and refrigerate.

Spicy Cinnamon Huckleberry Pecan Jam

Wow! This is a huckleberry jam with a spicy blend of pecans, ginger, and cinnamon to spread on toasted English muffins, crackers, or whatever you choose.

Ingredients:

- 4 c. fresh huckleberries, washed
- 1 c. chopped pecans
- 3 c. sugar
- 3 Tbs. brown sugar, firmly packed
- 1 tsp. ground ginger
- 1 Tbs. ground cinnamon
- 2 Tbs. apple cider vinegar
- 2 pkg. powdered pectin
- ¼ c. butter

Directions:

1. In large 6- to 8-quart nonreactive saucepan, combine huckleberries, pecans, sugars, ginger, cinnamon, and vinegar.
2. Cook over medium heat until sugar has dissolved and mixture is well blended.
3. Cook on simmer for 10 minutes; do not allow boiling.
4. Remove from heat; add butter to avoid foaming.
5. Pour into sterilized jars, and process following canning directions on page 186.

Yields: 2 half-pints.

Did You Know? . . . .

Did you know that the "Huckleberry," a camera periscope designed for Apple's MacBook computers, was first marketed in 2006?

Huckleberry Syrup

Try this delicious version of syrup made without corn syrup. It is full of true huckleberry flavor. While this recipe normally gives good results, fruit sugar, pectin, and acid concentrations can vary. Start with a small test batch and allow it to cool thoroughly before testing for syrup thickness.

Ingredients:

> 2 c. huckleberry juice
> 1¾ c. sugar
> 1 Tbs. lemon juice (optional) for a tarter syrup

Directions:

1. Crush fruit; press out juice using cheesecloth or jelly bag.
2. If you will not be making syrup immediately, pasteurize juice by heating to 194 degrees F. for 1 minute; filter through cheesecloth and refrigerate.
3. Mix juice and sugar in large pan, and bring to rolling boil that cannot be stirred down; continue to boil for 1 minute.
4. Remove pan from heat and skim off any foam.
5. Pour syrup into clean, hot canning jars.
6. Process following canning directions on page 186.
7. Refrigerate after opening.
8. This recipe produces fairly thin syrup. If you desire thicker syrup, use 1½ cups sugar and ¼ cup corn syrup in recipe.
9. Do not add more sugar or boil longer to thicken, because both methods can cause jelling to occur.
10. Corn syrup and lemon juice can be used together.

Did You Know? . . . .

Did you know that some of the best known huckleberries are native to the eastern and southeastern United States? They are from the genus Gaylussacia and are not found in the Western U.S.

Huckleberry Jam

This is a simple recipe with the pure taste of huckleberries for total delicious enjoyment.

Ingredients:

- 6   c. huckleberries
- 1   lemon, juiced, rind grated
- 7   c. sugar
- 1   bottle fruit pectin

Directions:

1. Rinse fruit thoroughly, and place in 6- to 8-quart nonreactive cooking pot.
2. Crush fruit; add lemon juice and grated rind of ½ lemon.
3. Add sugar; mix thoroughly.
4. Over medium-high heat, bring rapidly to full rolling boil, stirring constantly; boil hard 2 minutes.
5. Remove from heat and stir in fruit pectin; skim off foam.
6. Pour into sterilized jars, and process following canning directions on page 186.

Huckleberry-Bing Cherry Jam

This combination of two favorites, huckleberries and Bing cherries, makes a delicious jam. It is great on your favorite toast, bagel, or English muffin.

Ingredients:

- 3   c. huckleberries
- 2½ c. Bing cherries, pitted, chopped
- 2   Tbs. lemon juice
- 1   pkg. powdered fruit pectin
- 4   c. sugar
- 1   tsp. almond extract

Directions:

1. Purée huckleberries in blender.
2. In heavy pot combine huckleberries with chopped cherries, lemon juice, and pectin.
3. Bring to rolling boil over moderate heat, stirring constantly.
4. Add sugar all at once and bring mixture to boil again; continue stirring constantly.
5. Allow mixture to boil hard for 1 full minute.
6. Remove from heat, stir in almond extract, and skim off foam.
7. Pour into sterilized jars, and process following canning directions on page 186.

Yields: 6 to 7 half-pints.

Huckleberry Rhubarb Jam

This is a great way to use rhubarb, and the combination of the huckleberries and rhubarb makes a wonderfully tasty jam.

Ingredients:

1½ c. crushed huckleberries
¾ c. cooked, mashed rhubarb
3½ c. sugar
1   pouch liquid pectin (3 oz.)

Directions:

1. In large Dutch oven combine huckleberries, rhubarb, and sugar; mix well.
2. Place over high heat and bring to full rolling boil.
3. Cook for 1 minute, stirring constantly.
4. Remove from heat; stir in pectin.
5. Alternately stir and skim for 5 minutes.
6. Pour into sterilized jars, and process following canning directions on page 186.

Huckleberry Wine Jelly

This is one of those jellies that can be enjoyed on bread and butter or with your main meal as a complement to meats and vegetables.

Ingredients:

    3½ c. huckleberry wine (recipe page 293)
    ½   c. fresh lemon juice
    1   pkg. powdered pectin (2 oz.)
    4½ c. sugar

Directions:

1. Combine huckleberry wine, lemon juice, and pectin in large saucepot.
2. Bring to a boil, stirring frequently.
3. Add sugar, stirring until dissolved; return to rolling boil.
4. Boil hard 1 minute, stirring constantly; remove from heat and skim foam off top, if necessary.
5. Pour into sterilized jars, and process following canning directions on page 186.

Yields: 5 half-pints.

Red Currant and Huckleberry Jelly

This is a delicious jelly recipe from the 1940s, made with fresh currants and huckleberries.

Ingredients:

    2 lb. fresh huckleberries
    2 lb. fresh red currants
    1 c. water
    7 c. sugar
    4 fl. oz. liquid fruit pectin

Directions:

1. Place huckleberries and currants in large pot, and crush with potato masher or berry crusher.
2. Pour in water and bring to a boil; simmer for 10 minutes.
3. Remove from heat, cool, then strain fruit through jelly cloth or cheesecloth; measure out 5 cups of juice.
4. Pour juice into large saucepan and stir in sugar.
5. Bring to rapid boil over high heat, and stir in liquid pectin immediately; return to full rolling boil for 30 seconds.
6. Remove from heat and skim off foam from top.
7. Pour into sterilized jars, and process following canning directions on page 186.

Yields: 4 pints.

Hucklenana Jam

This could quickly become your family's favorite jam with its delicious, yummy taste!

Ingredients:

3 c. fresh huckleberries
1½ c. water
2 c. mashed bananas
7 c. sugar
6 fl. oz. liquid pectin
1 tsp. lemon juice

Directions:

1. In large saucepan combine huckleberries and water.
2. Place on medium heat and simmer for 10 minutes.
3. Stir in mashed bananas and sugar.
4. Increase heat to medium-high; boil 1 minute.
5. Remove from heat and stir in pectin and lemon juice; skim foam.
6. Pour into sterilized jars, and process following canning directions on page 186.

Spiced Huckleberry Jam

Some people enjoy the addition of spices to their jam; try this recipe with cinnamon and ginger added.

Ingredients:

    5   c. huckleberries
    1½ Tbs. lemon juice
    1   pkg. fruit pectin
    1   tsp. ground cinnamon
    ½   tsp. ground ginger
    4   c. sugar

Directions:

1. Place huckleberries in blender and process at low speed; transfer into heavy pot.
2. Stir in lemon juice, pectin, and spices; bring to a rolling boil over medium-high heat, stirring constantly.
3. Add sugar and bring to a boil again; boil for 1 minute.
4. Remove from heat and skim off foam.
5. Pour into sterilized jars, and process following canning directions on page 186.

Huckleberry Jam with Sherry

This version of huckleberry jam adds the flavor of orange and sherry for a delicious spread.

Ingredients:

    ½   gal. huckleberries, divided
    2   boxes fruit pectin
    2   pieces orange peel (4-in. lengths)
    7   c. sugar
    1   c. sherry
         juice from 2 oranges

Directions:

1. Wash and drain berries.

2. In blender, liquefy all but 1 cup of huckleberries together with orange juice.
3. Place in 6- to 8-quart pot; add remaining whole berries and stir in fruit pectin.
4. Add orange peel, and bring to rolling boil over moderately high heat, stirring constantly.
5. Remove from heat and add sugar and sherry.
6. Bring to boil again and cook 1 minute longer.
7. Remove from heat, skim off foam, and remove orange peel.
8. Pour into sterilized jars, and process following canning directions on page 186.

Yields: About 8 pints.

Huckleberry Preserves

This is a delicious blend of whole-berry preserves with great texture and flavor.

Ingredients:

6   c. huckleberries
1   Tbs. lemon juice
1   pkg. fruit pectin
4   c. sugar

Directions:

1. Rinse and drain huckleberries; place in bowl and mash with potato masher. (Should yield 4 cups.)
2. In heavy pan combine mashed huckleberries with lemon juice and pectin.
3. Bring to rolling boil over medium-high heat, stirring constantly.
4. Add sugar all at once, stirring constantly, and bring to a boil again; boil for 1 full minute.
5. Remove from heat, skim off foam, and ladle into sterilized hot jars.
6. Process following canning directions on page 186.

Yields: 6 to 7 half-pints.

Huckleberry Jalapeño Jelly

What better way to liven up sweet huckleberries than with the addition of spicy jalapeño peppers!

Ingredients:

1 c. huckleberries, fresh or frozen
½ c. chopped green bell pepper
¼ c. chopped jalapeño pepper
3 c. sugar
¾ c. apple cider vinegar
6 fl. oz. liquid pectin
1 sprig fresh mint, if desired

Directions:

1. In nonreactive 4-quart saucepan, combine huckleberries, bell pepper, and jalapeño pepper with sugar and cider vinegar.
2. Bring to a boil over medium-high heat, and boil rapidly for 1 minute; remove from heat and let stand for 5 minutes.
3. Stir in liquid pectin, then pour mixture through strainer to remove bits of peppers; skim off any foam.

Yields: 2 half-pints.

Did You Know? . . . .

Did you know that the cascade or blue huckleberry grows on Washington's Olympic Peninsula and in the Cascade Mountains from northern California into British Columbia? It grows at elevations between 1,900 and 6,600 feet in subalpine coniferous forests and alpine meadows.

Huckleberry Delights Cookbook

A Collection of Huckleberry Recipes
Cookbook Delights Series Book 6

Main Dishes

Table of Contents

Beef Medallions with Huckleberry Citrus Sauce

Beef medallions make an elegant dinner presentation. Try serving these huckleberry medallions with your favorite side dish of potato, pasta, or rice.

Ingredients:

- 1   c. huckleberries, fresh or frozen
- 6   Tbs. orange juice
- 6   Tbs. lemon juice
- 2   Tbs. vermouth
- 2   tsp. grated orange peel
- 2   tsp. grated lemon peel
- 1   tsp. minced fresh ginger
- 3   tsp. butter
- 1   lb. beef medallions
      salt and pepper to taste

Directions:

1. Combine huckleberries, orange juice, lemon juice, vermouth, orange peel, lemon peel, and ginger; stir to blend and set aside.
2. Melt butter in large, heavy skillet; sauté beef until brown and just cooked through.
3. Transfer beef to platter and keep warm.
4. Add huckleberry mixture to skillet; cook until mixture thickens, about 2 minutes, then salt and pepper to taste.
5. Spoon huckleberry mixture over beef and serve.

Yields: 3 servings.

Did You Know? . . . .

Did you know that researchers have identified 31 aromatic flavor chemicals in the fruits of the Cascade or blue huckleberry?

Beef with Huckleberries and Coconut

Serving over rice adds a complement to this delicious combination of beef and fruit with coconut.

Ingredients:

1	c. coarsely chopped yellow onions
2	Tbs. curry powder
2	tsp. salt
1	tsp. ground black pepper
½	c. vegetable oil
4	lb. boneless chuck, cut into 2-in. cubes
½	c. raisins
2	c. coconut milk
1	c. fresh huckleberries
1½	c. pineapple chunks
2	c. rice
5	c. salted water

Directions:

1. Sauté onions with curry powder, salt, and black pepper in vegetable oil until onions are soft.
2. Add beef cubes and cover pan immediately; do not allow meat to brown.
3. Simmer slowly for 1 hour.
4. Add raisins and coconut milk; simmer until meat is completely tender, about 1 hour.
5. Add huckleberries and pineapple chunks; cook for 10 minutes longer.
6. Adjust seasoning; sprinkle with more curry powder if desired.
7. Cook rice in boiling salted water; drain well.
8. Serve curry over rice in small bowls.

Yields: 12 servings.

Casablanca Salmon with Huckleberry Sauce

This is such a delicious way to serve salmon. It could easily become your family's favorite.

Ingredients for salmon:

1	c. orange juice
1	c. plain yogurt
1	c. water
4	salmon fillets (6 oz. each)

Ingredients for sauce:

1	Tbs. frozen orange juice concentrate, thawed
1	tsp. curry powder
¼	tsp. ground cinnamon
¼	tsp. ground nutmeg
¼	tsp. ground ginger
⅛	tsp. crushed red pepper
1	green onion, minced
2	Tbs. snipped cilantro
½	tsp. grated orange peel
1	clove garlic, crushed
4	Tbs. dried huckleberries (recipe page 238)
1	Tbs. sliced almonds
4	c. hot, cooked couscous
	cilantro sprigs for garnish

Directions for salmon:

1. Combine orange juice, yogurt, and water in large skillet for poaching liquid.
2. Bring to a boil and add salmon, then cover and reduce heat to simmer; simmer 8 to 10 minutes or until salmon is tender.
3. Remove salmon with slotted spoon to plate; cover and set aside.
4. Leave poaching liquid on simmer; reduce to about 1 cup.

Directions for sauce and assembly:

1. Combine orange juice concentrate, curry powder, ground spices, red pepper, and onion in bowl; mix well.

2. Add snipped cilantro, orange peel, garlic, huckleberries, and almonds; blend together and add to poaching liquid in skillet.
3. Stir and let simmer for about 5 minutes.
4. Place hot couscous on platter; arrange fillets on top.
5. Spoon a little sauce over top; serve remaining sauce in separate bowl on the side.
6. Garnish fish plate with cilantro sprigs.

Yields: 4 servings.

Trout with Huckleberry Stuffing

I was raised in Montana, and my brother used to fish for rainbow trout. My mom would cook them fresh, usually fried in cornmeal. This recipe is a very easy way to prepare trout with the addition of fresh Northwest huckleberries.

Ingredients:

2   tsp. butter, divided
1   lemon, sliced
1   whole trout, cleaned, or center-cut trout of choice
1   c. huckleberries
½   onion, chopped

Directions:

1. Spread 1 tsp. butter around center of piece of aluminum foil, arrange half of the lemon slices over butter, and place trout on top.
2. Mix together berries and onion; stuff fish.
3. Spread remaining butter on top of trout, arrange remaining lemon slices over top, fold and seal foil.
4. Grill 10 minutes per inch of trout at thickest point, turning every 5 to 10 minutes.
5. Unwrap foil, peel off skin, and remove bones.
6. Place trout fillets on plates; spoon stuffing and lemon slices on top.
7. Note: Substitute covered dish in place of foil, and bake in preheated oven at 350 degrees F. for 10 minutes per inch; follow same approach to serving.

Chicken and Huckleberry Angel Pasta

This easy and delicious, favorite family and company dish has a wonderful flavor.

Ingredients:

    ¼   c. butter
    1   pkt. Italian-style, dry salad dressing mix
    ½   c. white wine
    1   can condensed golden mushroom soup (10 oz.)
    4   oz. cream cheese with chives
    1¼ c. huckleberries, fresh or frozen
    6   skinless, boneless chicken breast halves
    1   lb. angel hair pasta

Directions:

1. Preheat oven to 325 degrees F.
2. In large saucepan melt butter over low heat; stir in dressing mix.
3. Blend in wine, mushroom soup, cream cheese, and huckleberries; heat through but do not boil.
4. Arrange chicken breasts in single layer in 13 x 9 x 2-inch baking dish; cover with sauce.
5. Cover with foil and bake for 60 minutes.
6. Twenty minutes before chicken is done, bring large pot of lightly salted water to a rolling boil.
7. Cook pasta until al dente, about 5 minutes; drain.
8. Place pasta on large serving platter and serve chicken with sauce over top.

Yields: 6 servings.

Did You Know? . . . .

Did you know that the red huckleberry, although somewhat sour, makes excellent pastries and preserves?

Sour Cream Soufflé with Huckleberries

Try this interesting soufflé. The Parmesan cheese and the sour cream diminish the sweetness and are a great contrast to the sweet huckleberries.

Ingredients:

6   lg. egg yolks
½   c. sour cream
¼   c. Parmesan cheese, grated
6   egg whites, beaten until stiff
3   Tbs. butter
½   c. sour cream, sweetened with 1 tsp. sugar or to taste
    fresh huckleberries

Directions:

1. Preheat oven to 325 degrees F.
2. Beat egg yolks until thick and lemon colored; about 5 minutes.
3. Beat in ½ cup of sour cream mixed with Parmesan cheese; fold in egg whites.
4. Melt butter in 10-inch, heavy ovenproof skillet.
5. Pour in egg mixture, leveling gently.
6. Cook over very low heat for 10 minutes.
7. Carefully move to preheated oven, and bake for 15 minutes until golden and puffed.
8. Remove from oven and immediately cut into 4 wedges; serve with dollop of the sweetened sour cream, and top with berries.

Yields: 4 servings.

Did You Know? . . . .

Did you know that the black or common huckleberry, dwarf huckleberry, and box huckleberry are all native to the eastern region of the United States?

Greek Huckleberry and Olive Chicken Pasta

This delicious pasta dish incorporates some of the flavors of Greece. It makes a wonderfully complete and satisfying meal.

Ingredients:

    1   lb. uncooked pasta
    1   Tbs. olive oil
    2   cloves garlic, crushed
    ½   c. chopped red onion
    1   lb. skinless, boneless chicken breast, cubed
    1   can artichoke hearts (14 oz.), drained, chopped
    2   c. fresh huckleberries, divided
    1   lg. tomato, chopped
    ½   c. crumbled feta cheese
    3   Tbs. chopped fresh parsley
    2   Tbs. lemon juice
    2   tsp. dried oregano
    ½   c. kalamata olives, chopped
    2   lemons, wedged, for garnish
        salt and pepper to taste

Directions:

1. Bring large pot of lightly salted water to a boil; cook pasta in boiling water for 8 to 10 minutes or until al dente; drain.
2. Meanwhile, heat olive oil in large skillet over medium-high heat; add garlic and onion, and sauté for 2 minutes.
3. Stir in chicken; cook, stirring occasionally, until chicken is no longer pink and juices run clear, about 5 to 6 minutes.
4. Reduce heat to medium-low, and add artichoke hearts, 1 cup huckleberries, tomato, feta cheese, parsley, lemon juice, oregano, and cooked pasta; stir until heated through, about 2 to 3 minutes.
5. Remove from heat; season to taste with salt and pepper.
6. Place in serving bowl, and sprinkle with remaining 1 cup huckleberries and kalamata olives; serve lemon wedges on the side.

Yields: 6 servings.

Huckleberry Chicken Breasts

This is another tasty and colorful chicken dish. Remember the importance of reducing the liquids in the sauce, thus obtaining the full flavor of the ingredients.

Ingredients:

    ½ tsp. Cajun spices, to taste
    4   boneless, skinless chicken breast halves
    3   tsp. olive oil
    3   cloves garlic, finely chopped
    1   med. onion, finely chopped
    ½ c. red wine
    2   c. huckleberries
        grated rind of 1 lemon
        salt and pepper to taste

Directions:

1. Dust chicken breasts with Cajun spices.
2. Sauté in olive oil until brown and almost cooked through, 7 to 10 minutes.
3. Remove chicken breasts from pan and keep warm.
4. In same pan, sauté garlic and onion until transparent, scraping remaining bits of chicken from bottom of pan.
5. Add red wine, huckleberries, and lemon rind; cook down, about 5 minutes, until most of liquid is evaporated.
6. Add salt and pepper to taste; let sit for 5 minutes, heat off, for flavors to blend.
7. Spoon over reserved chicken breasts and serve.

Yields: 4 servings.

Did You Know? . . . .

Did you know that the evergreen huckleberry is actually a species of blueberry and grows along the Pacific coast?

Pastrami Spiced Beef with Huckleberry Sauce

My family enjoys a spiced beef, and this marinated version is excellent served with huckleberry sauce. Remember that the trick is to serve the meat rich in spices.

Ingredients for spiced beef:

2	Tbs. coriander seeds
2	Tbs. mustard seeds
2	Tbs. white peppercorns
2	Tbs. allspice berries
6-8	dried red chilies
1	piece gingerroot (2-in. length), dried
1	piece cinnamon stick (2-in. length)
6	dried bay leaves
4	Tbs. kosher salt
1	c. huckleberries, fresh or frozen
4	Tbs. sugar
1	beef loin (4 lb.), fat trimmed
¼	c. peanut oil

Ingredients for huckleberry sauce:

2	Tbs. vegetable oil
1	lg. garlic clove, crushed
1	Tbs. minced fresh ginger
1	med. onion, finely chopped
2	pt. huckleberries, fresh or frozen
1	c. fresh tomato, peeled, seeded, chopped
2	lg. purple plums, pitted, chopped
¼	c. dark brown sugar, firmly packed
1	Tbs. huckleberry vinegar (recipe page 179)
1	Tbs. fresh lemon juice
1	med. dried chili pepper, crumbled
1	tsp. ground cinnamon
1	tsp. ground cardamom
1	tsp. ground coriander
1	tsp. salt
1	tsp. freshly ground mixed peppercorns
	zest of 1 lemon

Directions for spiced beef:

1. In spice mill or with mortar and pestle, grind spices, huckleberries, and sugar into paste, working in batches as necessary; combine batches of spices in small bowl.
2. Place beef loin on sheet of plastic wrap, and rub on all sides with pastrami spice paste; wrap tightly in plastic wrap to marinate.
3. Refrigerate wrapped loin for at least 12 hours and up to 24 hours.
4. Preheat oven to 400 degrees F.
5. In large sauté or roasting pan, heat vegetable oil over high heat.
6. Add beef to pan and sear quickly on all sides, being careful not to burn spices.
7. Place seared beef in oven and bake about 45 minutes or until thermometer inserted into center reads 142 degrees F.
8. Remove loin from oven; cover and let rest for 15 minutes.
9. Carve and serve with huckleberry sauce. (Recipe follows.)

Directions for huckleberry sauce:

1. Heat oil in heavy-bottomed, 2-quart saucepan; add garlic and ginger, and cook over low heat for 2 minutes.
2. Add onion; cook until soft and transparent, stirring often.
3. Add huckleberries, tomato, plums, brown sugar, vinegar, lemon juice, zest, chili pepper, spices, salt, and pepper; stir well.
4. Cook over medium heat until mixture begins to simmer.
5. Reduce heat; continue simmering gently for 30 minutes.
6. Remove from heat and let mixture cool slightly; purée in food processor or blender.
7. Return purée to pan and heat, bringing mixture to a simmer; cook until thick, about 1 hour.
8. Pour amount not using into sterile half-pint jars or containers; cover and let cool.
9. Store in refrigerator for up to 4 weeks or may be frozen up to 4 months.

Yields: 6 servings.

Roasted Chicken with Huckleberry Sauce

Huckleberry sauce adds a delicious flavor to this roasted chicken that the family loves.

Ingredients:

 2 c. port wine
 2 c. water
 ¾ c. sugar, divided
 ¼ c. honey
 1 whole chicken (4 lb.), wing tips removed
 2 Tbs. olive oil
 ⅓ c. diced shallots
 ½ c. cider vinegar
 ¼ c. brandy
 1 c. fresh huckleberries
 1½ tsp. chopped fresh tarragon
 2 c. chicken stock
    salt and ground black pepper
    fresh huckleberries for garnish
    carrot rosettes for garnish

Directions:

1. In pot large enough to hold chicken, combine port, water, ¼ cup sugar, and honey; bring to a boil.
2. Meanwhile, season chicken cavity with salt and black pepper; truss chicken and set aside.
3. When wine mixture comes to a boil, remove from heat.
4. Add chicken; let marinate at room temperature for 30 minutes, turning occasionally.
5. Preheat oven to 350 degrees F.
6. Remove chicken from marinade; place on rack in shallow roasting pan.
7. Bake for about 1 hour and 15 minutes or until chicken is brown and reaches internal temperature of 185 degrees F.
8. Remove from oven and let stand at room temperature for 10 minutes.

9. Meanwhile, in medium skillet heat olive oil until hot.
10. Add shallots; cook, stirring frequently, until soft, 3 to 5 minutes.
11. Add remaining ½ cup sugar; cook over low heat, stirring frequently, until sugar is dark brown, 10 to 15 minutes.
12. Stir in vinegar, brandy, huckleberries, and tarragon.
13. Cook over medium-high heat until huckleberries are soft, about 5 minutes.
14. Add chicken stock; bring to a boil, and boil until reduced by half, about 20 minutes.
15. Season to taste with salt and pepper.
16. To serve, carve chicken, dividing among serving plates.
17. Spoon huckleberry sauce over chicken.
18. Garnish with fresh huckleberries, carrot roses, etc., if desired.

Yields: 2 to 4 servings with 1½ cups sauce.

Huckleberry Filling for Pierogies

Huckleberries make a great filling for old-fashioned pierogies.

Ingredients:

    4   c. fresh huckleberries
    3   tsp. sugar

Directions:

1. Wash berries and drain.
2. In medium bowl sprinkle berries with sugar.
3. Mix lightly.
4. Fill pierogi shells and process immediately, before juice is drawn out of the fruit.

Yields: 4 cups, enough for 40 to 45 pierogies.

Peppercorn-Crusted Muscovy Duck with Huckleberries

This is a great recipe for special occasions. Muscovy duck breasts, which are larger and meatier than other ducks, can be found at specialty stores.

Ingredients:

¼   c. sugar
2   Tbs. water
2½ Tbs. balsamic vinegar
1½ c. low-salt chicken broth
1   c. mixed dried fruit, cut into matchstick-size pieces
1   Tbs. minced fresh ginger
2   boneless Muscovy duck breast halves with skin (12 to 16 oz. each)
4   tsp. crushed mixed peppercorns
¼   c. port
2   Tbs. chilled butter, cut into small pieces
¾   c. frozen huckleberries, thawed

Directions:

1. Stir sugar and water in heavy, small saucepan over low heat until sugar dissolves.
2. Increase heat; boil without stirring until mixture is deep amber, occasionally brushing down sides of pan with wet pastry brush and swirling pan, about 8 minutes.
3. Stir in vinegar (mixture will bubble); add broth.
4. Simmer until reduced to 1 cup, about 20 minutes; remove from heat.
5. Stir in dried fruit and ginger; let stand 30 minutes.
6. Meanwhile, using fork, pierce duck skin (not meat) all over; sprinkle with salt.
7. Rub pepper over skin side of duck.
8. Place duck, skin side down, in cold cast-iron skillet; turn on heat to medium.
9. Cook until golden and crisp, about 15 minutes.
10. Turn over; cook to desired doneness, about 8 minutes for medium; let rest 10 minutes.

11. Rewarm sauce over low heat; stir in port.
12. Whisk in butter a few pieces at a time; stir in huckleberries.
13. Season with salt and pepper.
14. Slice duck breasts; serve with sauce.

Yields: 4 servings.

Salmon with Huckleberry Horseradish Sauce

Salmon, huckleberries, and horseradish combine to make a piquant Northwest dish.

Ingredients:

1	lb. salmon fillet
1	c. water
1	pt. fresh huckleberries
½	c. sugar
½	lemon, juiced
1	oz. freshly grated horseradish
	fresh basil sprigs for garnish

Directions:

1. Cut salmon into six 2-ounce pieces, and poach in skillet until just underdone or internal temperature reaches 125 degrees F.
2. Remove with slotted spoon and cool immediately in refrigerator.
3. Bring water to a boil and add huckleberries.
4. Reduce heat and simmer just until tender.
5. Add sugar and cook for 2 more minutes.
6. Finish cooking another minute with lemon juice and grated horseradish.
7. Remove from heat; stir in basil.
8. Spoon pool onto plate and serve warm with chilled salmon on top.
9. Garnish with a sprig of basil.

Yields: 2 to 3 servings.

Shrimp in Curried Huckleberry Sauce

If you like both seafood and curry, try this absolutely delightful dish for a change of pace.

Ingredients:

- 1   Tbs. olive oil
- 1   Tbs. unsalted butter
- ½   clove garlic, pressed, or ¼ tsp. garlic powder
- 1   lb. raw fresh shrimp, peeled, cleaned
- 4   Vidalia sweet onions, cut into eighths
- 1½ c. fresh huckleberries
- 2   Tbs. curry powder (Madras preferred)
- ¼   tsp. cayenne pepper
- 1-2 c. water
-      salt and pepper to taste

Directions:

1. Heat large skillet to medium heat and add olive oil, butter, and garlic; after butter melts, add shrimp and reduce heat to low.
2. Cook shrimp on one side, about 2 minutes; turn to cook on other side for 1 minute.
3. Transfer shrimp to bowl, leaving juices in skillet.
4. Add onions, huckleberries, curry powder, and cayenne pepper; stir then add water and stir again.
5. Cover, and simmer onions for 20 to 30 minutes on low; salt and pepper to taste
6. Add cooked shrimp; cook uncovered for 3 minutes.
7. Remove from heat and serve while hot.

Yields: 2 servings.

Did You Know? . . . .

Did you know that the word "huckleberry" is first found in American English around the year 1670? It was most likely a variation of "whortleberry."

Huckleberry Delights Cookbook

A Collection of Huckleberry Recipes
Cookbook Delights Series Book 6

Pies

Table of Contents

Page

A Basic Recipe for Pie Crust

This is a very good recipe for a delicious, flaky crust.

Ingredients for single crust:

> 1½ c. sifted all-purpose flour
> ½   tsp. salt
> ½   c. shortening
> 4-5 Tbs. ice water

Ingredients for double crust:

> 2   c. sifted all-purpose flour
> 1   tsp. salt
> ⅔   c. shortening
> 5-7 Tbs. ice water

Directions for single crust:

1. In large bowl stir together flour and salt.
2. Cut in shortening with pastry blender or mix with fingertips until pieces are size of coarse crumbs.
3. Sprinkle 2 tablespoons ice water over flour mixture, tossing with fork.
4. Add just enough remaining water 1 tablespoon at a time to moisten dough, tossing so dough holds together.
5. Roll pastry into 11-inch circle and wrap in plastic wrap; refrigerate for 1 hour.
6. Preheat oven to 425 degrees F.
7. Remove plastic wrap from pastry, and fit pastry into a 9-inch pie plate.
8. Fold edge under and then crimp between thumb and forefinger to make fluted crust.
9. For filled pie with an instant or cooked filling (cream-filled, custard-filled, etc.), prick crust all over with fork then bake 15 to 20 minutes until done.
10. If preparing pie with uncooked filling (such as pumpkin), do not prick crust; pour filling into unbaked pastry shell, and then bake as directed.

Directions for double crust:

1. Turn desired filling into pastry-lined pie plate; trim overhanging edge of pastry ½ inch from rim of plate.
2. Cut slits with knife in top crust for steam vents.
3. Place over filling; trim overhanging edge of pastry 1 inch from rim of plate.
4. Fold and roll top edge under lower edge, pressing on rim to seal; flute.
5. Cover fluted edge with 2- to 3-inch-wide strip of aluminum foil to prevent excessive browning.
6. Remove foil during last 15 minutes of baking.

Yields: 1 pie crust (9-inch single or double).

A Basic Cookie or Graham Cracker Crust

This is a great crust for use with cream pies or for an unbaked pie. Use your favorite flavor of cookie to complement your filling, or use graham crackers.

Ingredients:

2   c. cookie or graham cracker crumbs, finely crushed
⅓   c. sugar
½   c. butter, melted

Directions:

1. Combine crumbs, butter, and sugar.
2. Press mixture firmly against bottom and up sides of 9-inch pie plate.
3. Baking is not necessary, but if preferred crust may be baked at 400 degrees F. for 10 minutes.

Yields: 1 pie crust (9-inch).

Huckleberry Coconut Pie

This huckleberry pie keeps well in your freezer for surprise guests or for your invited company.

Ingredients:

1   egg, well beaten
1¼ c. flaked coconut
¼   c. chopped walnuts
¼   c. light corn syrup
1   Tbs. all-purpose flour
¼   c. plus ⅔ c. sugar, divided
1   pkg. frozen huckleberries (10 oz.), unsweetened
1   c. heavy cream, whipped
1   unbaked, 9-inch single-crust pastry shell (recipe page 216)

Directions:

1. Preheat oven to 375 degrees F.
2. Bake crust for 5 minutes then remove from oven; set aside and reduce oven temperature to 350 degrees F.
3. In medium bowl combine egg with coconut, nuts, syrup, flour, and ¼ cup sugar; spread in bottom of partly baked pastry shell.
4. Return pie to oven and bake for 15 minutes; remove to wire rack and cool thoroughly.
5. Crush frozen huckleberries and combine with remaining ⅔ cup sugar; fold into whipped cream.
6. Pour berry mixture over cooled coconut mixture and freeze.
7. When ready to serve, thaw pie slightly and cut into wedges.

Yields: 6 to 8 servings.

Did You Know? . . . .

Did you know that it can take 10 to 15 years for a burned huckleberry field to return to full productivity?

Wild Huckleberry Pie

This is a very good huckleberry pie, and it is not quite as sweet as some. Use fresh or frozen huckleberries. This one uses tapioca as a thickening agent. It is good to try all versions of the pie to see which one you like the best.

Ingredients:

- 6 c. huckleberries, fresh or frozen
- 3 Tbs. tapioca
- ⅔ c. sugar
- ½ c. brown sugar, firmly packed
- 1 Tbs. cider vinegar
- 1 Tbs. butter
- 1 unbaked double-crust pastry shell (recipe page 216)

Directions:

1. Preheat oven to 400 degrees F.
2. In large mixing bowl combine huckleberries, tapioca, sugar, brown sugar, and vinegar.
3. Pour mixture into unbaked pastry shell; dot top with butter.
4. Add top pastry and flute edges.
5. Bake for 15 minutes; reduce oven to 350 degrees F., and bake for 45 to 55 minutes, adding another 15 minutes if berries are frozen.
6. Remove from oven, and place on wire rack to cool before slicing to serve.

Yields: 6 to 8 servings.

Did You Know? . . . .

Did you know that several small butterflies, such as the early spring flying Brown Elfin and Henry's Elfin, occasionally lay their eggs on the huckleberry plant?

Huckleberry Cream Cheese Pie

This is a refreshing and attractive, chilled huckleberry pie. It is so creamy and delicious!

Ingredients:

1   baked single-crust pastry shell (recipe page 216)
3   oz. cream cheese, softened
½   c. powdered sugar
1   tsp. vanilla extract
2   c. fresh huckleberries
1   c. sugar
½   c. water
3   Tbs. cornstarch
½   pt. whipping cream, sweetened to taste
    fresh huckleberries, for garnish

Directions:

1. Prepare pastry shell; set aside.
2. Mix cream cheese, powdered sugar, and vanilla together.
3. Spread cream cheese mixture onto bottom of baked pie shell.
4. In saucepan combine huckleberries, sugar, water, and cornstarch.
5. Bring to a boil, stirring constantly; cook until thickened.
6. Pour over cream mixture in pie shell, and cool completely.
7. Whip sweetened cream and spread on top of cooled pie.
8. Garnish with fresh huckleberries; cut into wedges and serve.

Yields: 6 to 8 servings.

Did You Know? . . . .

Did you know that there is a low spreading shrub in the southwestern United States called the huckleberry oak that has small acorns and leaves that resemble the huckleberry?

Latticed Huckleberry Pie

This makes an attractive huckleberry pie with the lattice top. There are no spices added to this so you can enjoy the taste of pure huckleberries.

Ingredients:

- 5   c. huckleberries
- 2   tsp. almond extract
- 3½ Tbs. all-purpose flour
- 1¼ c. sugar
- ½   tsp. salt
- 1   unbaked double-crust pastry shell (recipe page 216)

Directions:

1. Preheat oven to 375 degrees F.
2. Prepare pastry and divide in half; roll out first crust, line pie pan, and set aside.
3. Sprinkle extract over berries.
4. Mix flour, sugar, and salt; combine with huckleberries and transfer to crust-lined pie pan.
5. Roll out second crust, and cut strips about ½ inch wide with pastry wheel or knife.
6. Place half of strips horizontally over filling.
7. Lay second set of strips vertically across first layer; weave strips if desired.
8. Trim ends of strips; fold and flute edge of pie, building a higher crust to prevent filling from bubbling over.
9. Loosely cover with foil.
10. Bake for about 20 minutes; remove foil and bake another 25 minutes.
11. Allow to cool on wire rack before slicing to serve.

Yields: 6 to 8 servings.

Huckleberry Marshmallow Pie

This is an attractive, easy-to-make pie that tastes absolutely great! This pie can be made ahead and kept ready to serve for unexpected guests.

Ingredients:

- 32 lg. marshmallows
- ½ c. half-and-half
- 1 c. whipped cream
- 2 c. huckleberries, fresh or frozen
- 1 baked single-crust pastry shell or graham cracker crust shell (recipe page 216)

Directions:

1. Prepare baked pastry shell or graham cracker crust; set aside.
2. In double boiler melt marshmallows with half-and-half; cool to lukewarm.
3. Fold whipped cream into marshmallow mixture.
4. Gently fold huckleberries into marshmallow and cream mixture.
5. Spoon into baked pie shell or graham cracker crust pie shell and freeze.
6. When ready to serve, unthaw slightly and cut into wedges.

Yields: 6 to 8 servings.

Did You Know? . . . .

Did you know that there is a box huckleberry plant in Pennsylvania that has been classified as the oldest living thing? It is more than 13,000 years old and covers an area about ¼ mile in diameter.

Sour Cream Huckleberry Pie

The pecan topping on this pie adds a delightful flavor and crunch.

Ingredients for filling:

1	unbaked 9-in. single-crust pie shell (recipe page 216)
1	c. sour cream
2	Tbs. all-purpose flour
¾	c. sugar
1	tsp. vanilla extract
¼	tsp. salt
1	egg, beaten
2½	c. huckleberries, fresh or frozen

Ingredients for pecan topping:

3	Tbs. all-purpose flour
3	Tbs. butter, softened
3	Tbs. chopped pecans

Directions for filling:

1. Preheat oven to 400 degrees F.
2. In mixing bowl beat together sour cream, flour, sugar, vanilla, salt, and egg until smooth, about 4 to 5 minutes.
3. Gently fold in huckleberries.
4. Pour into pie crust and bake for 25 minutes.
5. Remove from oven, and add topping as described below.

Directions for pecan topping:

1. Combine flour, butter, and pecans, mixing well.
2. Sprinkle pecan mixture over top of pie; return to oven and bake 10 minutes longer.
3. Let cool, then chill before serving.

Yields: 8 to 10 servings.

Huckleberry Pie Squares

This is a unique version of huckleberry pie recipes and makes a tasty combination.

Ingredients for crust:

2½ c. all-purpose flour
1   Tbs. sugar
½   tsp. salt
1   c. butter-flavor shortening
2   eggs, separated
¼   c. milk
1   tsp. lemon juice

Ingredients for filling:

¾   c. corn flakes, finely crushed
6½ c. frozen huckleberries, unthawed
1¼ c. sugar
½   tsp. ground cinnamon

Ingredients for glaze:

2   egg whites
1   c. powdered sugar
2   Tbs. milk
½   tsp. vanilla extract
¼   tsp. almond extract
    Sweetened Whipped Cream (recipe page 167)

Directions for crust:

1. Combine flour, sugar, and salt; cut in shortening.
2. Mix egg yolks with milk and lemon juice; add to flour mixture, stirring until dough clings together.
3. Divide dough in half; chill for 1 hour.
4. Roll out between two sheets of wax paper into 16 × 12-inch rectangles.
5. Fit one sheet evenly into 15 x 10 x 1-inch jellyroll pan; reserve second rolled out sheet.

Directions for filling:

1. In medium bowl combine corn flakes and huckleberries.

2. Add sugar and cinnamon; toss well.
3. Place on rolled out rectangle on jellyroll pan to ½ inch of edge.
4. Cover with second sheet of dough; dampen and seal edges with fork.

Directions for glaze:

1. Preheat oven to 375 degrees F.
2. Beat egg whites lightly, and brush some of it over pastry.
3. Bake for 55 to 60 minutes.
4. Remove from oven and cool on wire rack to lukewarm.
5. Mix powdered sugar, milk, vanilla, and almond extract together until well blended; drizzle over warm crust.
6. Cool completely before cutting into 2-inch squares.
7. Serve each square with a spoonful of whipped cream over the top.

Yields: 2 dozen squares.

Lemon Huckleberry Pie

The flavor of lemon makes this pie perfectly delicious.

Ingredients:

1	recipe double-crust pastry (recipe page 216)
1¼	c. sugar
⅓	c. all-purpose flour
¼	tsp. salt
½	tsp. ground cinnamon
2	Tbs. lemon juice
2	tsp. lemon zest
5	c. fresh huckleberries

Directions:

1. Preheat oven to 425 degrees F.
2. Prepare pastry and line pie plate with half of dough.
3. Gently mix together all ingredients; pour into pie crust.
4. Cover with remaining dough and gently press edges to seal.
5. Prick top several times with fork to allow steam to escape, and bake for 40 minutes.

Huckleberry Pie with Phyllo Crust

Phyllo always tastes great, and this makes a change of pace for your favorite huckleberry pie.

Ingredients for phyllo crust:

- 2 tsp. sugar
- 1 tsp. all-purpose flour
- 6 phyllo sheets
  melted butter

Ingredients for filling:

- ¾ c. sugar
- 3 Tbs. cornstarch
- 1 tsp. ground cinnamon
- 6 c. huckleberries
- 1½ Tbs. fresh lemon juice

Directions for phyllo crust:

1. Lightly coat 9-inch pie pan with melted butter.
2. In small bowl combine sugar and flour; set aside.
3. Stack 5 phyllo sheets and cut them in half crosswise; reserve 1 sheet for lattice top.
4. Cover sheets with slightly damp cloth until ready to use.
5. Take one sheet and layer it in the pan.
6. Coat lightly with melted butter and sprinkle with ½ teaspoon of sugar mixture.
7. Repeat with remaining sheets, layering clockwise at 1-inch intervals until entire pie pan rim is covered.
8. Trim excess phyllo with kitchen shears; set aside.

Directions for filling:

1. In 4-quart saucepan combine sugar, cornstarch, and cinnamon; mix well.
2. Add huckleberries and sprinkle with lemon juice; toss lightly to combine.
3. Cook and gently stir over medium heat until mixture comes to a boil.
4. Cook an additional 2 minutes, stirring constantly.

5. Remove from heat; cool to lukewarm and transfer to prepared crust.

Directions for lattice top and baking:

1. Preheat oven to 350 degrees F.
2. Cut reserved phyllo sheet in half crosswise.
3. Lightly coat each half sheet with nonstick vegetable oil spray.
4. Stack the 2 layers and fold lengthwise.
5. Cut into 6 strips.
6. Twist strips and arrange them over pie filling–3 horizontal and 3 vertical strips.
7. Bake for 20 to 25 minutes or until phyllo is golden.
8. Remove from oven; cool on wire rack.
9. Slice and serve at room temperature.

Yields: 6 to 8 servings.

Huckleberry Cheese Pie

Huckleberries are also great with cream cheese and sour cream. Try this excellent variation of huckleberry pie.

Ingredients:

1   graham cracker crust (recipe page 217)
2   pkg. cream cheese (8 oz. each)
½   c. sour cream
2   Tbs. sugar
1   c. huckleberry pie filling (recipe page 237 or 239)

Directions:

1. Prepare graham cracker crust; set aside.
2. Preheat oven to 350 degrees F.
3. In medium bowl blend cream cheese, sour cream, and sugar until smooth; pour into pie shell.
4. Top with huckleberry pie filling.
5. Sprinkle sugar on top and bake for 5 minutes.
6. Remove from oven; cool on wire rack, then chill and serve.

Huckleberry Cream Pie

This is another version of a huckleberry and cream pie. The cream makes an excellent combination with the flavor of huckleberries.

Ingredients:

1 single-crust pastry shell, unbaked (recipe page 216)
3 eggs, beaten
1 c. sugar
1 dash salt
2 Tbs. all-purpose flour
5 c. huckleberries
⅓ c. whipping cream

Directions:

1. Prepare pastry shell; set aside.
2. Preheat oven to 375 degrees F.
3. Combine eggs, sugar, salt, and flour; stir in huckleberries.
4. Pour into unbaked pastry shell.
5. Slowly pour whipping cream over top, being sure to cover entire area.
6. Bake for 10 minutes.
7. Reduce heat to 350 degrees F., and bake an additional 50 minutes, until filling is set.
8. Serve hot or cold with ice cream or whipped cream.

Yields: 8 servings.

Did You Know? . . . .

Did you know that some people call the red-black huckleberry the "southern cranberry"?

Apple Huckleberry Pie

This makes a delicious combination of apples and huckleberries, always best served warm with homemade vanilla or cinnamon ice cream.

Ingredients:

- ¾   c. sugar
- 3   Tbs. cornstarch
- 5   c. apples, peeled, sliced
- 2½ c. fresh huckleberries
- 1   Tbs. lemon juice
- 2   Tbs. butter
        pastry for double-crust pie (recipe page 216)

Directions:

1. Preheat oven to 425 degrees F.
2. In large bowl stir together sugar and cornstarch.
3. Add apples, huckleberries, and lemon juice; toss to coat fruit.
4. Turn into pastry-lined 9-inch pie plate; dot with butter.
5. Add top crust, seal and flute edge, and cut slits in top crust for steam vents.
6. Bake for 1 hour or until crust is browned and filling is bubbly.
7. Remove from oven, and cool on wire rack before slicing to serve.

Yields: 6 to 8 servings.

Did You Know? . . . .

Did you know that the dwarf huckleberry, generally only 1 to 2 feet tall, bears large quantities of black huckleberries that are ¼ to ⅓ inch in diameter?

Huckleberry-Orange Pie with Orange Crust

Orange and huckleberry combine to make a very smooth-tasting version of huckleberry pie.

Ingredients for orange pastry crust:

1¼ c. all-purpose flour
1   tsp. finely grated orange zest
1   Tbs. sugar
½   tsp. salt
½   c. butter-flavored shortening
1½ Tbs. reconstituted orange juice
1½ Tbs. cold water

Ingredients for filling:

¾   c. sugar
3   Tbs. cornstarch
⅓   c. frozen orange juice concentrate, thawed
¼   c. reconstituted orange juice
7   c. fresh huckleberries, rinsed, well drained, divided

Directions for orange pastry crust:

1. Combine flour, orange zest, sugar, and salt in mixing bowl; stir to blend.
2. Using pastry blender or two knives, cut shortening into flour mixture until it resembles coarse meal.
3. Combine orange juice with cold water.
4. Add juice/water by tablespoons, mixing gently with fork just until dough begins to hold together in clumps. (If necessary an extra tablespoon of water may be added.)
5. Gather dough and shape in flat round disk.
6. Chill at least 1 hour.
7. On lightly floured wax paper, roll dough to about ⅛ inch thick, in big enough circle to overhang pie plate by 1½ to 2 inches. (A quick, accurate measure can be attained by turning pie plate upside down onto pastry and then adding 2 inches.)

8. Ease pastry into pie plate, being careful not to stretch dough, and fold pastry under at edge.
9. Crimp decoratively; prick bottom and sides.
10. Freeze briefly, 15 to 20 minutes.
11. Preheat oven to 375 degrees F.
12. Bake for 15 minutes or until golden brown.
13. Cool on wire rack.

Directions for filling:

1. Stir together sugar and cornstarch in heavy-bottomed, 3-quart saucepan.
2. Stir in orange juice concentrate, orange juice, and 2 cups of huckleberries.
3. Cook and stir over medium-high heat until mixture is thickened, translucent, and just comes to a boil, 7 to 10 minutes.
4. Remove from heat and gently fold in remaining huckleberries.
5. Mound filling into baked orange pastry crust.
6. Refrigerate at least one hour before serving.
7. Can be made one day ahead.

Yields: 8 to 10 servings.

Did You Know? . . . .

Did you know that the unripe fruits of the garden huckleberry are toxic and bitter, but the fully ripe fruits are edible when cooked?

Did you know that wild berries, including wild huckleberries, are among the ceremonial foods served at the First Salmon Celebration, a festival celebrated each spring by the Pacific Northwest Native Americans for centuries?

Huckleberry Ginger Ice Box Pie

This refrigerated huckleberry pie is tart and refreshing on hot summer days.

Ingredients:

1	graham cracker crust, 10 in. (recipe page 217)
5	c. fresh huckleberries, divided
½	c. sugar
1	knob fresh ginger (1-in. length), grated
¼	c. cornstarch
¼	c. water
1	Tbs. butter
	juice of ½ lemon
	zest of ½ lemon

Directions:

1. In medium saucepan bring 3 cups huckleberries, sugar, ginger, and lemon juice and zest to a boil.
2. Meanwhile, whisk together cornstarch and water, set aside.
3. After huckleberries come to a boil, stir in cornstarch mixture and cook 2 additional minutes.
4. Remove from heat and stir in remaining 2 cups of huckleberries and butter.
5. Pour into prepared pie shell.
6. Refrigerate for at least 6 to 8 hours or up to 1 day before serving.

Yields: 8 to 10 servings.

Did You Know? . . . .

Did you know that the berries from the evergreen huckleberry plant are said to be even tastier after freezing?

Huckleberry Delights Cookbook

A Collection of Huckleberry Recipes
Cookbook Delights Series Book 6

Preserving

Table of Contents

A Basic Guide for Canning, Dehydrating, and Freezing

1. Place empty jars in hot, soapy water. Wash well inside and out with brush or soft cloth.
2. Run your finger around rim of each jar, discarding any that are chipped or cracked.
3. Rinse in clean, clear, very hot water, being careful to use tongs to avoid burning skin or fingers.
4. Place upside down on towel or fabric to drain well.
5. Place lids in boiling water bath for 2 minutes to sterilize and keep hot until ready to place on jar rims.
6. Immediately prior to filling jars with hot food, immerse in hot bath for 1 minute to heat jars. Heating jars avoids breakage.
7. If filling with room-temperature food, you need not immerse immediately prior to filling.
8. Fill jars with food to within ½ inch of neck of jars.
9. When ladling liquid over food, fill jars to 1 inch from top rim in each jar. This leaves air allowance for sealing purposes.
10. Wipe rims of jars with damp, clean cloth to remove any particles of food and again check for chips or cracks.
11. Using tongs, place lids from hot bath directly onto jars.
12. Place rings over lids, and using cloth, gloves, or holders, tighten down firmly while hanging onto jars.
13. Do not tighten down too hard as air may become trapped in jars and prevent them from sealing.
14. For fruits, tomatoes, and pickled vegetables, place each jar into water bath canning kettle so water covers jars by at least 1 inch.
15. For vegetables, process them in a pressure canner according to manufacturer's directions.
16. Follow time recommended for food being canned.
17. Do not mix jars of food in same canning kettle as times may vary for each kind of food.
18. At end of time recommended for canning, gently lift each jar out of bath with tongs, and place on protected surface.
19. Turn lids gently to be sure they are firmly tight.
20. Place filled, ringed jars on cloth to cool gradually.
21. Do not disturb rings, lids, or jars until sealed.

22. Lids will show slight indentation when sealed.
23. When cool, wipe jars with damp cloth then label and date each jar.
24. Leave overnight until thoroughly cooled.
25. Jars may then be stored upright on shelves.

Dehydrating

1. Always begin with fresh, good quality food that is clean and inspected for damage.
2. Pretreatment is not necessary, but food that is blanched will keep its color and flavor better. Use the same blanching times as you would for freezing. Fruit, especially, responds well to pretreatment.
3. Doing some research on pretreatments may help you decide what procedure you would like to use.
4. You can marinate, salt, sweeten, or spice foods before you dehydrate them.
5. Jerky is meat that has been marinated and/or flavored by rubbing spices into it; avoid oil or grease of any kind as it will turn rancid as the food dries.
6. Vegetables and fruit can be treated the same way.
7. Slice or dice food thin and uniform so that it will dehydrate evenly. Uneven thicknesses may cause food to spoil because it did not dry as thoroughly as other parts.
8. Space food on dehydrator tray so that air can move around each piece.
9. Try not to let any piece touch another.
10. Fill your trays with all the same type of food as different foods take different amounts of time to dry.
11. You can, of course, dry different types of food at the same time, but you will have to remember to watch and remove the food that dehydrates more quickly. You can mix different foods in the same dehydrator batch, but do not mix strong vegetables like onions and garlic as other foods will absorb their taste while they are dehydrating.
12. The smaller the pieces, the faster a food will dehydrate. Thin leaves of spinach, celery, etc., will dry fastest. Remove

them from the stalks before drying them or they will be overdone, losing flavor and quality. In very warm areas, they might even scorch. If they do, they will taste just like burned food when you rehydrate them.

13. Dense food like carrots will feel very hard when they are ready. Others will be crispy. Usually, a food that is high in fructose (sugar) will be leathery when it is finished dehydrating.
14. Remember that food smells when it is in the process of drying, so outdoors or in the garage is an excellent place to dry a big batch of those onions!
15. Always test each batch to make sure it is "done."
16. You can pasteurize finished food by putting it in a slow oven (150 degrees F.) for a few minutes.
17. Let the food cool before storing.
18. Store in airtight containers to guard against moisture. Jars saved from other food work well as long as they have lids that will keep moisture out.
19. Zip-closure food storage bags work well.
20. Jars of dehydrated carrots, celery, beets, etc., may look cheerful on your countertop, but the colors and flavors will fade. Dehydrated food keeps its color and flavor best if stored in a dark, cool place.
21. Dehydrating food takes time, so do not rush it. When you are all done, you will have a dried food stash to be proud of!

Freezing

1. Wash all containers and lids in hot, soapy water using soft cloth.
2. Rinse well in clear, clean, hot water.
3. Cool and drain well.
4. Place food into container to within 1 inch of rim. This allows for expansion of food during freezing.
5. Wipe rim of container with clean damp cloth, checking for chips or breaks.
6. Be certain cover fits the container snugly to avoid leaks. Burp air from container.
7. If food is hot when placing in container, cool prior to placing in freezer.
8. Label and date each container.
9. Store upright in freezer until frozen solid.

Canned Huckleberry Pie Filling

This recipe is for one quart of pie filling. Just multiply amounts by however many quarts you desire to make.

Ingredients:

> 3½ c. huckleberries, fresh or frozen, thawed
> ¾   c. plus 2 Tbs. sugar
> ¼   c. plus 1 Tbs. pectin gel
> 1   c. cold water
> ½   c. lemon juice
> 3   drops blue food coloring (optional)
> 1   drop red food coloring (optional)

Directions:

1. Wash and drain fresh huckleberries.
2. For fresh fruit, place 6 cups at a time in 1 gallon boiling water.
3. Boil each batch 1 minute after water returns to a boil.
4. Drain, but keep heated fruit in covered bowl or pot.
5. Combine sugar and pectin gel in large kettle; stir.
6. Add water and, if desired, food coloring.
7. Cook on medium-high heat until mixture thickens and begins to bubble.
8. Add lemon juice and boil 1 minute, stirring constantly.
9. Fold in drained berries immediately, and fill jars with mixture without delay, leaving 1-inch headspace.
10. Adjust lids and process immediately canning directions on page 234.
11. Processing times for pints or quarts are 30 minutes for 1 to 1,000 feet, 35 minutes for 1,001 to 3,000 feet, 40 minutes for 3,001 to 6,000 feet, and 45 minutes above 6,000 feet.
12. After processing is complete, turn off heat and remove canner lid.
13. Wait 5 minutes before removing jars.

Yields: 1 quart.

Dried Huckleberries

Here is a good way to keep those huckleberries you worked so hard to pick. It is nice to enjoy them all year round, and it is a convenient way of storing them.

Ingredients:

  huckleberries, rinsed
  lemon juice

Directions:

1. Rinse huckleberries and let sit in lemon juice for 1 hour.
2. Make small cut or hole in each berry; spread out to dry on paper towels and wire rack about 2 to 3 days or until roughly consistency of raisins.
3. May also dry in food dehydrator using manufacturer's instructions, on a screen in a gas oven with only the pilot light on, or hang them with a needle and thread.
4. Store in airtight container for up to 1 month.

Freezing Huckleberries

Make sure you use only ripe, full-flavored huckleberries to freeze for the best results.

Ingredients:

  huckleberries

Directions for whole berries:

1. Sort huckleberries; wash quickly in cool water only if very dirty, otherwise freeze as is.
2. Pat dry with paper towels so they will not stick together when freezing.

3. Spread berries in single layer on metal tray; freeze until solid.
4. When frozen, pack in containers and label.
5. To freeze with syrup, pack in containers and cover with 40 percent syrup (3 cups sugar to 4 cups water).

Directions for purée:

1. Wash berries and purée in blender or food processor.
2. Mix 1 cup sugar into each quart of puréed berries.
3. Stir until sugar is dissolved.
4. Pack into containers, leaving 1 inch of headspace.

Frozen Huckleberry Pie Filling

Freezing your pie filling makes it quick and easy to make pies at a later date.

Ingredients:

12 c. huckleberries
3   c. sugar
¾   c. cornstarch
1   Tbs. grated lemon peel
¼   c. lemon juice

Directions:

1. Wash and drain huckleberries.
2. Combine sugar and cornstarch in saucepan.
3. Stir huckleberries into sugar mixture and let stand until juice begins to flow from berries, about 30 minutes.
4. Add lemon peel and juice.
5. Cook over medium heat until mixture begins to thicken.
6. Ladle filling into freezer containers or bags, leaving ½-inch headspace.
7. Cool to room temperature for 2 hours before freezing.

Yields: 5 pints.

Huckleberry Fruit Leather

My children love all kinds of fruit leather, and huckleberry fruit leather makes an extra special treat. Adults will also love this treat, and it is a great choice for company.

Ingredients:

    6   c. ripe huckleberries
    1½ c. apple juice

Directions:

1. Combine berries and juice in 2-quart saucepan; cook over moderate heat until mixture boils.
2. Lower heat and cook a few minutes longer, until berries are soft.
3. Remove from heat, then cool to lukewarm and mash.
4. Strain through fine sieve or double layer of cheesecloth, pressing juice from pulp; discard pulp.
5. Return juice to saucepan and bring to a boil; stir and cook over moderate heat until reduced in volume by half and does not run.
6. Remove from heat and cool; spread 4-inch circles evenly on wax paper-covered cookie sheet.
7. Carefully place in oven at 200 degrees F., and dry 4 to 6 hours with door ajar to reach leather-like consistency; return to oven for additional time if necessary.
8. Or, if using food dehydrator, follow manufacturer's directions.
9. Roll and wrap separately; store in airtight container.

Did You Know? . . . .

Did you know that the red huckleberry, a species native to the Pacific Northwest of North America, is common in forests from southeastern Alaska and British Columbia south through Washington and Oregon to central California?

Huckleberry Lemon Jam

This jam tastes great on English muffins or even biscuits.

Ingredients:

    4½ c. huckleberries, fresh or frozen
    7   c. sugar
    2   Tbs. lemon juice
    3   pouches liquid pectin (3 oz. each)
        grated zest of 2 large lemons

Directions:

1. Pick over fresh huckleberries to remove any stalks, and rinse under cold water.
2. Drain well and place in large, heavy-bottomed saucepan. (Do not rinse or thaw frozen berries.)
3. Crush berries slightly with potato masher or pestle.
4. Stir in sugar, lemon juice, and zest.
5. Bring to a boil over medium-high heat, stirring often.
6. When mixture reaches full boil, cook for 1 minute.
7. Stir in pectin; return to full boil, then cook for 1 minute more.
8. Ladle into hot, sterilized jars leaving ¼ inch of headspace.
9. Wipe rims clean and put lids on top of jars.
10. Process in boiling water bath for about 5 minutes, following canning directions on page 234.
11. Remove from water and cool completely at room temperature.

Yields: About 6 cups.

Did You Know? . . . .

Did you know that the rating for the evergreen huckleberry as feed for livestock is fair to poor for sheep, goats, and deer; poor to useless for cattle; and useless for horses?

Huckleberry Juice

This is a great way to preserve some of your huckleberries so you have it available year round to use in recipes calling for juice.

Ingredients:

huckleberries

Directions:

1. Sort and wash berries.
2. Place in large cooking pot and cover with water.
3. Cook just below boiling point for 30 minutes.
4. Strain through double thickness of cheesecloth; discard pulp.
5. Can as is or add 1 cup sugar to each gallon of juice for a smooth drink.
6. Pour hot juice into sterilized jars, and process following canning directions on page 234.

Huckleberry Chutney

Chutneys are great over poultry or other meats. This chutney is very colorful and full of flavor.

Ingredients:

4	c. huckleberries, fresh or frozen
1	can whole-berry cranberry sauce (16 oz.)
¼	c. sugar
3	Tbs. balsamic vinegar
1½	tsp. grated orange peel
1	tsp. ground ginger
¼	tsp. crushed red pepper
¼	tsp. ground black pepper

Directions:

1. In medium, nonreactive saucepan combine huckleberries, cranberry sauce, sugar, balsamic vinegar, orange peel, ginger, and red and black peppers.
2. Bring to a boil; boil uncovered, stirring frequently, until slightly thickened, 15 to 20 minutes.
3. Pour into sterilized jars, and process following canning directions on page 234.

Yields: 5 cups.

Freezer Huckleberry Jam

This is an easy freezer huckleberry jam with pectin.

Ingredients:

3   c. huckleberries, rinsed well
1   Tbs. lemon juice
5¼ c. sugar
1   pkg. powdered pectin
¾   c. water

Directions:

1. Crush berries in large bowl; add lemon juice.
2. Stir in sugar and let stand for 10 minutes.
3. In saucepan mix pectin and water; bring to a full boil.
4. Boil, stirring constantly, for 1 minute.
5. Add pectin mixture to fruit; stir for 3 minutes.
6. Pour fruit mixture into freezer containers or small canning jars, leaving ½-inch headspace.
7. Cover immediately and let stand at room temperature for 24 hours.
8. Store in freezer.

Yields: About 6 cups.

Huckleberry Orange Vinegar

This is a wonderful and flavorful vinegar to use in salads, sauces, and anywhere you would need vinegar. It also makes a nice gift when put in decorative bottles.

Ingredients:

    2 pt. huckleberries, washed, drained
    1 c. sugar
       vinegar to cover
       zest of 1 orange

Directions:

1. Place huckleberries in large bowl.
2. Cover with vinegar and let stand for 1 hour.
3. Transfer to large, nonreactive saucepan.
4. Add sugar and orange zest; bring to a boil.
5. Reduce heat; cover and simmer for 20 minutes.
6. Strain through fine sieve or double layer of cheesecloth, pressing out as much liquid as possible.
7. Discard residue and pour liquid back into pan.
8. Heat through and pour into hot glass jars or fancy bottles.
9. Refrigerate after opening.

Huckleberry Lime Jam

The addition of lime to this jam makes it tart and refreshing.

Ingredients:

    4½ c. huckleberries
    1 pkg. powdered pectin
    5 c. sugar
    1 Tbs. grated lime peel
    ⅓ c. lime juice

Directions:

1. Crush berries one layer at a time.
2. Combine crushed huckleberries and pectin in large, heavy-bottomed pan.
3. Bring to a boil, stirring frequently.
4. Add sugar, stirring until dissolved.
5. Stir in grated lime peel and lime juice.
6. Bring to a rolling boil; boil hard 1 minute, stirring constantly.
7. Remove from heat; skim foam if necessary.
8. Ladle hot jam into hot jars, leaving ¼-inch headspace; adjust two-piece caps.
9. Process in boiling water bath for about 15 minutes, following canning directions on page 234.

Yields: About 6 half-pints.

Huckleberry-Banana Leather

This fruit leather makes a delicious snack that can be packed in lunch boxes or eaten at any time.

Ingredients:

1½ c. diced bananas
1½ c. huckleberries
2   Tbs. fresh lemon juice

Directions:

1. Purée all ingredients.
2. Spread 4-inch circles evenly on wax paper-covered cookie sheet.
3. Carefully place in oven at 200 degrees F., and dry 4 to 6 hours with door ajar to reach leather-like consistency; return to oven for additional time if necessary.
4. Or, if using food dehydrator, follow manufacturer's directions.
5. Roll and wrap separately; store in airtight container.

Huckleberry Marmalade

My children love marmalade, and this version is delicious.

Ingredients:

1	med. orange
1	med. lemon
¾	c. water
⅛	tsp. baking soda
4	c. fresh huckleberries, crushed
5	c. sugar
1	pkg. liquid fruit pectin (6 oz.)

Directions:

1. Peel orange and lemon; finely chop rind and place in large cooking pan.
2. Chop orange and lemon pulp; set aside.
3. Add ¾ cup water and baking soda to rind and bring to boil.
4. Reduce heat and simmer 10 minutes, stirring occasionally.
5. Add chopped orange and lemon pulp, huckleberries, and sugar; return to a boil.
6. Reduce heat and simmer 5 minutes; remove from heat and cool 5 minutes.
7. Add pectin; return to a boil.
8. Boil, stirring constantly for 1 minute; remove from heat and skim off foam with metal spoon.
9. Pour into hot sterilized jars, filling to ¼ inch from top; wipe jar rims, and cover at once with metal lids and screw on bands.
10. Process in boiling water bath for about 10 minutes, following canning directions on page 234.

Yields: 6 half-pints.

Triple-Berry Vinegar

This vinegar can be used as a marinade and can be brushed over fish or chicken before grilling. Or you can try adding a splash of it to fruit juice or carbonated water.

Ingredients:

- ¼ c. fresh huckleberries
- ¼ c. fresh raspberries
- ¼ c. frozen cranberries, thawed
- 8 sage leaves
- 1 c. white wine vinegar
- 3 whole cloves
- 2 cinnamon sticks (3-in. lengths)
- 1 Tbs. sugar
- 8 black peppercorns
- 5 fresh huckleberries
- 5 fresh raspberries
- 5 frozen cranberries
- 4 sage leaves

Directions:

1. Combine first 4 ingredients in nonreactive bowl; crush with spoon.
2. Place berry mixture, vinegar, cloves, and cinnamon in wide-mouthed jar; cover and let stand 2 weeks in cool, dark place, gently shaking jar occasionally.
3. Strain vinegar mixture through sieve lined with cheesecloth into glass measure or medium bowl; discard solids.
4. Pour strained vinegar into small, nonreactive saucepan; add sugar.
5. Cook 5 minutes over low heat or until sugar dissolves; cool.
6. Pour into decorative bottle; add peppercorns and remaining ingredients.
7. Seal with cork or other airtight lid.

Canned Huckleberries

It is always a good idea to stock up on huckleberries when they are in season to enjoy year round.

Ingredients:

- 1   gal. fresh huckleberries
- 4   c. water
- 3   c. sugar

Directions:

1. Rinse and drain huckleberries.
2. Place in sterilized pint jars.
3. In saucepan bring water and sugar to boil over medium-high heat, stirring constantly.
4. Pour hot mixture over huckleberries, leaving ⅜-inch headspace.
5. Process following canning directions on page 234.

Yields: 8 pints.

Huckleberry Balsamic Vinegar

Our family loves balsamic vinegar. This is very flavorful and makes a great salad dressing, marinade, or sauce.

Ingredients:

- 4   c. huckleberries, fresh or frozen, thawed
- 1   qt. balsamic vinegar
- ¼   c. sugar
- 1   lime, peeled, cut into strips (green part only)
- 1   cinnamon stick

Directions:

1. In large, nonreactive saucepan crush huckleberries with potato masher or back of heavy spoon.
2. Add vinegar, sugar, lime, and cinnamon; bring to a boil.
3. Reduce heat and simmer, covered, for 20 minutes.

4. Cool slightly and pour into large bowl.
5. Cover and refrigerate for 2 days to allow flavors to blend.
6. Place wire mesh strainer (or cheesecloth) over large bowl.
7. In batches, ladle huckleberry mixture into strainer, pressing out as much liquid as possible; discard solids.
8. Pour vinegar into clean glass bottles or jars; refrigerate, tightly covered, indefinitely.
9. Use in salad dressings or drizzled over grilled chicken or beef.

Yields: 5 cups.

Huckleberry Pineapple Relish

This is a delicious relish that is a great accompaniment to any meal. Just wait until your family gets a taste of this one! You will be making more very soon!

Ingredients:

2   c. fresh huckleberries, chopped
1   c. diced fresh pineapple
1   lg. jalapeño pepper, seeded, finely chopped
2   Tbs. fresh lime juice
1   Tbs. chopped fresh cilantro leaves
2   tsp. light brown sugar, firmly packed
⅛   tsp. salt

Directions:

1. In processor, pulse berries to coarsely chop; repeat separately to coarsely chop pineapple, then combine both in medium mixing bowl.
2. Add jalapeño, lime juice, cilantro, brown sugar, and salt; blend well.
3. Refrigerate up to 4 hours before eating if using fresh.
4. For preserving, bring to a boil in 2-quart saucepan; turn heat to simmer and cook for 8 minutes.
5. Pour into sterilized jars, and process following canning directions on page 234.

Huckleberry Topping

This huckleberry topping can be used on anything from pancakes to vanilla ice cream. It is best served warm, right after it is made, but it also works well to can or freeze it for later use.

Ingredients:

- 2 qt. huckleberries, rinsed, drained
- 2 c. water
- 1 Tbs. grated lemon peel
- 3 c. sugar
- 4 c. water
- 2 Tbs. lemon juice

Directions:

1. Place huckleberries in large stockpot and crush with potato masher.
2. Add water and lemon peel; simmer over low heat for 5 minutes.
3. Strain fruit through cheesecloth, metal strainer, or food mill to remove pulp and seeds; reserve juice.
4. In large saucepan mix sugar and 4 cups water.
5. Bring to a boil, stirring occasionally, until temperature reaches 260 degrees F.
6. Add huckleberry juice and boil for 1 minute.
7. Stir in lemon juice and remove from heat.
8. Let cool and freeze or can as desired, following canning directions on page 234.

Yields: 6 cups.

Did You Know? . . . .

Did you know that the larva of the Huckleberry Sphinx moth feeds on huckleberry and blueberry foliage as well as cherry and willow?

Huckleberry Delights Cookbook

A Collection of Huckleberry Recipes
Cookbook Delights Series Book 6

Salads

Table of Contents

Fruit Rice Salad with Huckleberries

The combination of the berries with almonds and whipped cream is so delicious that everyone will want more. Be sure to make extra!

Ingredients:

1	can crushed pineapple (15 oz.), drained
½	c. strawberries, sliced
¼	c. fresh peaches, sliced
1¼	c. cooked white rice, chilled
⅓	c. golden raisins
⅔	c. flaked coconut
¾	c. heavy whipping cream
2	Tbs. sugar
¼	tsp. vanilla extract
⅛	tsp. ground ginger
¼	c. slivered almonds
12	leaves lettuce
½	c. sliced almonds for garnish
1	c. fresh huckleberries for garnish

Directions:

1. In medium bowl mix together pineapple, strawberries, peaches, rice, raisins, and coconut; set aside.
2. In separate large bowl whip together cream, sugar, vanilla, and ginger until stiff.
3. Fold in rice mixture then stir in slivered almonds.
4. Spoon mixture onto lettuce leaf beds.
5. Garnish each portion with sliced almonds and fresh huckleberries.

Yields: 6 servings.

Huckleberries, Mushrooms, and Hazelnut Salad

This makes a great-tasting Northwest salad and an attractive presentation for your expertise!

Ingredients:

- 4 c. huckleberries
- 1 c. sugar
- 2 c. water
- 2 c. raspberry vinegar
- 2 c. plus 1 Tbs. olive oil, divided
- 4 oz. Chanterelle mushrooms
- 8 oz. mixed greens
- 3 oz. hazelnuts
  crumbled goat cheese, your favorite, for garnish

Directions:

1. Bring huckleberries, sugar, and water to a boil in saucepan; reduce by half.
2. Combine 1 cup reduced huckleberries, raspberry vinegar, and 2 cups olive oil; blend well.
3. Sear Chanterelle mushrooms in remaining 1 tablespoon olive oil until soft.
4. Cover mixed greens with warm mushrooms and hazelnuts.
5. Add 2 ounces warm vinaigrette dressing.
6. Garnish with goat cheese.

Yields: 4 servings.

Did You Know? . . . .

Did you know that the black huckleberry is the only kind found in the Chicago region?

Huckleberries in Plum Wine

Fresh berries tossed with a simple dressing of wine and lime juice make a refreshing and colorful salad.

Ingredients:

> 1   c. fresh huckleberries
> 1   c. fresh raspberries
> ¼   c. plum wine or other sweet, fruity wine
> 1½ tsp. lime juice
> 1½ tsp. sugar
>   favorite salad greens

Directions:

1. In medium bowl combine berries, wine, lime juice, and sugar.
2. Toss well and refrigerate until ready to serve.
3. Serve, chilled, on bed of your favorite salad greens.

Yields: 4 servings.

Huckleberry Ambrosia Salad

There are many different versions of ambrosia. Try this version with colorful huckleberries.

Ingredients:

> 1   can pineapple chunks, drained, chopped
> 1   can sliced peaches (16 oz.), drained, chopped
> 1   can mandarin oranges (11 oz.), drained, chopped
> 3   c. whipped cream
> 1   c. shredded coconut
> ½   c. chopped walnuts
> 2   c. huckleberries
> ¼   c. sugar

Directions:

1. In large bowl combine pineapple, peaches, and mandarin oranges with whipped cream.
2. Gently fold in coconut and walnuts.
3. Toss huckleberries with sugar and fold 1 cup into salad; chill until ready to serve.
4. Place in large serving bowl or on individual serving plates with a lettuce leaf if desired; sprinkle remaining huckleberries over top.

Yields: 8 to 10 servings.

Huckleberry Banana Kiwi Fruit Salad

Bananas, strawberries, kiwis, and peaches make such a bright combination in this simple, yet great-tasting, fruit salad that it is perfect for your summertime celebrations.

Ingredients:

1   c. huckleberries
12 lg. strawberries, sliced
2   med. kiwis, sliced
1½ peaches, sliced
2   med. bananas, sliced
    lemon juice
    honey
    lettuce leaves

Directions:

1. In large bowl combine huckleberries, strawberries, kiwis, and peaches; toss together lightly.
2. Mix bananas and lemon juice together; add to other fruits.
3. Sweeten to taste with honey and chill for up to 1 hour.
4. Serve in large bowl or on individual plates with lettuce leaf base if desired.

Yields: 6 servings.

Huckleberry and Chicken Salad

With the huckleberries this makes an attractive and colorful chicken salad that is great served over your favorite greens.

Ingredients:

¼ c. sour cream
½ c. mayonnaise
2 c. cooked chicken breasts, cubed
½ c. green onion or scallions, sliced
¾ c. celery, diagonally sliced
½ c. sweet red bell pepper, diced
1½ c. fresh or frozen huckleberries
    choice of greens
    fresh or frozen huckleberries for garnish
    lemon slices for garnish

Directions:

1. In medium bowl combine sour cream and mayonnaise.
2. Add chicken, green onion, celery, and bell pepper; mix gently.
3. Fold in huckleberries.
4. Cover and refrigerate at least 30 minutes to let flavors blend.
5. Serve on bed of endive or other greens.
6. Garnish with huckleberries and lemon slices, if desired.

Yields: 6 servings.

Did You Know? . . . .

Did you know that the Kwakwaka'wakw people of British Columbia cooked huckleberries with salmon roe, and the Sechelt people smoke-dried them, using huckleberry branches as part of the fuel?

Minted Huckleberry Melon Salad

Always choose the best fruits and berries. Your salad will only be as good as the fruit you select.

Ingredients:

 1 c. water
 ¾ c. sugar
 3 Tbs. lime juice
 1½ tsp. fresh mint, chopped
 ¾ tsp. aniseed
 5 c. watermelon, cubed
 3 c. honeydew, cubed
 3 c. cantaloupe, cubed
 2 c. peach slices
 1 c. fresh huckleberries

Directions:

1. In small saucepan combine water, sugar, lime juice, mint, and aniseed; bring ingredients to a boil.
2. Boil for 2 minutes; remove from heat.
3. Cover and cool syrup completely; set aside.
4. Combine watermelon, honeydew, cantaloupe, peaches, and huckleberries in large bowl.
5. Add syrup and stir lightly to coat.
6. Cover and chill in refrigerator for at least 2 hours, stirring occasionally.
7. Remove from refrigerator and drain well before serving.

Did You Know? . . . .

Did you know that huckleberries are important to wildlife for cover, browse, and fruits? Deer browse on the twigs and foliage, and wildlife such as bears, ruffed grouse, quail, turkey, and squirrels eat the fruits.

Huckleberry and Apple Compote

This compote is very easy to make. Serve it as a light, healthy salad with or without lettuce leaves.

Ingredients:

- 1   c. apples, unpeeled, cored, cubed
- 2   tsp. fresh lime juice
- 1   c. huckleberries
- ½   c. dried huckleberries (recipe page 238)
- ½   c. wild hickory nuts, coarse chopped, lightly toasted
  lettuce leaves if desired

Directions:

1. In medium bowl combine apples with lime juice; toss well.
2. Add berries and nuts; toss lightly.
3. Chill until ready to serve, up to 1 hour.
4. Serve on individual plates with lettuce leaf base if desired.

Yields: 4 servings.

Creamy Huckleberry Salad

This is a unique salad that really highlights the flavor of the huckleberries.

Ingredients:

- 1   lg. pkg. black cherry gelatin
- 2   c. boiling water
- 1   c. plain yogurt
- 1   c. sour cream
- ¼   tsp. ground ginger
- 2   tsp. grated lemon rind
- 2   c. fresh huckleberries
  mint sprigs for garnish
  fresh strawberries for garnish

Directions:

1. In large bowl mix gelatin and boiling water, stirring until gelatin dissolves.
2. Smoothly combine yogurt, sour cream, ginger, and lemon rind in separate bowl; add to gelatin mixture, blending well.
3. Refrigerate, stirring occasionally, until mixture is slightly thickened and syrupy.
4. Remove from refrigerator; fold in huckleberries.
5. Pour into 6-cup mold; refrigerate 4 to 6 hours until firm.
6. Unmold onto serving plate; garnish with mint and fresh strawberries.

Yields: 6 to 8 servings.

Huckleberry Cantaloupe Nectarine Fruit Salad

Always use the freshest, top-quality fruits available for the best taste. Enjoy this simple and colorful combination salad with family or company!

Ingredients:

3 firm peaches or nectarines, skins slipped, sliced
1 sm. cantaloupe, peeled, cubed
½ c. raspberries
½ c. huckleberries
¾ c. frozen orange juice concentrate, thawed
   lettuce leaves

Directions:

1. In large bowl combine peaches or nectarines, cantaloupe, and berries.
2. Pour orange juice concentrate over fruit, and toss lightly to mix; chill up to 1 hour.
3. When ready to serve, scoop onto individual serving plates with bed of lettuce leaves.

Yields: 6 to 8 servings.

Huckleberry Walnut Salad

Here is a perfect salad for any season! It is so delicious with the combination of berries, nuts, and greens. For an entrée, add chicken, diced apples, and diced green onions.

Ingredients:

- 1   pkg. mixed salad greens (10 oz.)
- 1   pt. fresh huckleberries
- ¼   c. walnuts
- ½   c. raspberry vinaigrette salad dressing
- ¼   c. crumbled feta cheese

Directions:

1. In large bowl combine salad greens with huckleberries and walnuts.
2. Toss with raspberry vinaigrette.
3. Top with feta cheese to serve.

Yields: 6 servings.

Wild Rice Huckleberry Salad

This salad is easy to make and tastes wonderful. Wild rice is much healthier than white rice, and the taste is delicious!

Ingredients:

- 1   pkg. wild rice (6 oz.)
- ¾   c. light mayonnaise
- 1   tsp. white vinegar
- 1   tsp. sugar
- 1   c. dried huckleberries (recipe page 238)
- ¼   c. green onion, diced
- 1   c. seedless red grapes, halved
- 8   oz. blanched slivered almonds
    salt and pepper to taste

Directions:

1. Cook rice according to package directions.
2. Remove from heat and set aside to cool.
3. In medium bowl whisk together mayonnaise, vinegar, sugar, then salt and pepper to taste.
4. Stir in rice, dried huckleberries, onion, and grapes until evenly coated with dressing.
5. Cover and refrigerate for 1 to 2 hours.
6. Before serving in large bowl or on individual plates, sprinkle slivered almonds over top of salad.

Huckleberry Potato Salad

This is an unusual combination, but it actually makes a very good potato salad. Your family will like it as well as ours does.

Ingredients:

¼ c. white wine vinegar
1   Tbs. olive oil
½   tsp. sugar
½   tsp. salt
½   tsp. dried basil, crushed
⅛   tsp. black pepper
4   lg. red potatoes, cooked, sliced
1   c. fresh huckleberries
½   c. cucumber, diced
½   c. carrot, shredded
2   Tbs. scallions, chopped, for garnish
2   Tbs. parsley for garnish

Directions:

1. In shaker jar combine vinegar, oil, sugar, salt, basil, and pepper; blend well.
2. In large bowl combine dressing with potatoes, mixing well.
3. Add huckleberries, cucumber, and carrot; toss well.
4. Serve in large bowl or on individual plates; sprinkle with chopped scallions and parsley.

Huckleberry Rice Salad

This tasty rice salad goes great with tangy lamb kebabs.

Ingredients:

- 3 c. cooked rice, room temperature
- ½ c. sliced almonds, toasted
- 1 c. fresh huckleberries
- ½ c. vinaigrette

Directions:

1. To toast sliced almonds, preheat oven to 350 degrees F.; place almonds in single layer on baking sheet.
2. Place in middle of oven for approximately 5 minutes or until lightly toasted. (Watch carefully because they burn easily.)
3. In medium bowl combine rice, almonds, and huckleberries; lightly toss to combine.
4. Stir vinaigrette and measure ½ cup.
5. Add vinaigrette to rice mixture and lightly toss.
6. Taste; add a little more vinaigrette if necessary.

Cantaloupe with Chicken Salad

This makes a delightful presentation for your guests to enjoy.

Ingredients:

- 2 c. cubed, cooked chicken
- 1¾ c. fresh huckleberries
- 1 c. sliced celery
- 1 c. seedless green grapes
- ½ c. sliced almonds
- 3 cantaloupes, halved, seeded
- ½ c. mayonnaise
- ¼ c. sour cream
- 1 Tbs. fresh lemon juice
- 1½ tsp. grated lemon peel
- 1½ tsp. sugar

½ tsp. ground ginger
¼ tsp. salt (optional)

Directions:

1. In large bowl combine chicken, huckleberries, celery, grapes, and almonds.
2. In small bowl mix remaining ingredients except cantaloupe.
3. Pour over chicken mixture and toss gently.
4. Spoon into cantaloupe halves and serve.

Yields: 6 servings.

Huckleberry Green Salad Vinaigrette

This is a great way to get your family to eat their fruits and vegetables and enjoy them at the same time!

Ingredients for dressing:

3 Tbs. sugar
2 Tbs. apple cider vinegar
¼ tsp. dry mustard
¼ c. olive oil

Ingredients for salad:

20 oz. fresh spinach, stems removed, washed, drained
1 c. fresh huckleberries
3 oranges, peeled, sectioned, cut in thirds
½ c. pecans or slivered almonds

Directions:

1. Combine sugar, vinegar, mustard, and oil in shaker jar; set aside.
2. Tear spinach into bite-size pieces, and place in salad bowl.
3. Add huckleberries, oranges, and nuts to spinach leaves.
4. Pour prepared dressing over salad; toss lightly to mix.
5. Serve individually or in large bowl.

Fresh Spinach and Fruit Salad

The sweet-tart taste of the salad dressing on this salad complements the bitterness of the spinach and greens.

Ingredients for salad:

- 5 c. fresh spinach, washed, trimmed
- 5 c. mixed field greens
- 1 c. strawberries, cut in half (or 1 c. orange slices)
- 1 Granny Smith apple, cored, sliced
- ¼ c. orange juice
- ½ c. dried huckleberries (recipe page 238)
- ½ med. red onion, thinly sliced
- ½ c. Monterey Jack cheese, shredded
- ½ c. honey mustard vinaigrette (recipe follows)
- ¼ c. nuts or seeds (optional)

Ingredients for vinaigrette:

- ¼ c. olive oil
- 2 Tbs. red wine vinegar
- 1 Tbs. Dijon mustard
- 1 Tbs. honey
- ½ tsp. salt
- ¼ tsp. pepper
- ⅛ tsp. ground nutmeg

Directions for salad:

1. Toss sliced apples in orange juice to coat to prevent browning.
2. Toss all salad ingredients together.
3. Drizzle vinaigrette over salad and lightly toss again.
4. Top with nuts or seeds if desired.

Directions for vinaigrette:

1. Whisk together all ingredients.
2. Cover and chill up to 4 days.

Yields: 10 servings of salad; ½ cup vinaigrette.

Huckleberry Green Salad

Our family loves bleu cheese, and this is a very flavorful and colorful combination of huckleberries and bleu cheese on top of your favorite greens.

Ingredients for dressing:

⅓ c. orange juice
1 Tbs. lemon juice
1 Tbs. white wine vinegar
1 garlic clove, minced
3 Tbs. huckleberry jam (recipe page 192)
¼ tsp. dry mustard
¼ tsp. paprika
¼ tsp. dried oregano
¼ tsp. dried thyme

Ingredients for salad:

8 c. romaine lettuce, torn to bite size
¼ c. slivered almonds, toasted
⅓ c. crumbled bleu cheese
⅓ c. whole huckleberries

Directions:

1. In shaker jar combine orange and lemon juices, vinegar, garlic, huckleberry jam, and spices; shake well.
2. Toss romaine lettuce with desired amount of dressing; divide among 4 individual chilled plates.
3. Sprinkle toasted almonds and bleu cheese over each salad.
4. Top with fresh whole huckleberries.

Yields: 4 servings.

Fruit Salad with Amaretto Cream Sauce

This salad is started the day before serving, so you will need to plan ahead.

Ingredients:

- ¼ pt. raspberries
- ¼ pt. huckleberries
- ¼ pt. strawberries, washed, halved
- ¼ c. diced orange segments
- ¼ c. diced grapefruit segments
- ¼ c. diced Granny Smith apple
- ¼ c. halved seedless green grapes
- ¼ c. diced peach
- ¼ c. diced apricot
- ¼ c. lemon juice
- ¼ c. plus 2 Tbs. sugar, divided
- 1 tsp. chopped fresh mint
- ¼ c. heavy cream
- 2 egg yolks
- ¼ c. Amaretto liqueur

Directions:

1. Mix together all fruit, lemon juice, ¼ cup sugar, and mint.
2. Cover and refrigerate overnight.
3. The next day, bring cream to a boil then set pan aside to cool slightly.
4. Whisk egg yolks and remaining 2 tablespoons sugar together.
5. When cream has cooled, whisk in egg and sugar mixture.
6. Strain sauce and stir in liqueur.
7. Serve in small pitcher to pour over individual salad servings.

Yields: 4 servings.

Huckleberry Delights Cookbook

A Collection of Huckleberry Recipes
Cookbook Delights Series Book 6

Side Dishes

Table of Contents

Blintz Soufflé with Huckleberries

Here is a soufflé that will complement the wonderful taste of any main dish you want to put on the table!

Ingredients:

8	oz. cream cheese, softened
2	c. small curd cottage cheese
2	egg yolks
⅓	c. plus 1 Tbs. sugar, divided
1	tsp. vanilla extract
6	whole eggs
1½	c. sour cream
½	c. orange juice
½	c. butter, softened
1	c. all-purpose flour
2	tsp. baking powder
1	tsp. grated orange rind
1½	c. fresh huckleberries

Directions:

1. Preheat oven to 350 degrees F.
2. In large bowl cream together cream cheese, cottage cheese, egg yolks, 1 tablespoon sugar, and vanilla; blend well.
3. In separate bowl beat together whole eggs, sour cream, orange juice, and butter.
4. Sift together flour, ⅓ cup sugar, and baking powder; add to above egg and juice ingredients; stir in orange rind and blend until smooth.
5. Pour half of batter into greased 13 x 9 x 2-inch baking pan; spoon cream cheese mixture over batter and smooth with knife.
6. Pour remaining batter on top of cream cheese layer; sprinkle huckleberries over top.
7. Bake for 50 to 60 minutes, until puffy and golden.
8. Remove from oven and serve immediately.

Huckleberry Alla Checca

Here is a wonderful recipe for an uncooked sauce made with fresh huckleberries and garden tomatoes. You do not have to cook a thing except for the pasta! What a delicious side dish this makes to complement your meal.

Ingredients:

1½	c. fresh huckleberries
5	med. tomatoes, seeded, diced, drained
4	cloves garlic, minced
½	c. chopped fresh basil
½	c. olive oil
½	c. red wine
1	lb. pasta
	grated Parmesan cheese

Directions:

1. Combine huckleberries, tomatoes, garlic, basil, olive oil, and wine in nonmetal bowl.
2. Cover with plastic wrap; allow to sit at room temperature at least 2 hours or as long as 10 hours to marinate flavors.
3. Cook pasta in large pot of boiling salted water until al dente; drain well.
4. Pour uncooked sauce over hot pasta; toss.
5. Serve with grated Parmesan cheese on the side.

Yields: 4 servings.

Did You Know? . . . .

Did you know that after their huckleberry feast, the native Sahaptin people of the Columbia Plateau would migrate to the mountains to gather berries and to escape the summer heat?

Huckleberries and Fresh Fruit Conserve

Try serving this delightful conserve with pork or chicken for tonight's delicious dinner.

Ingredients:

- ½ c. water
- 2 c. fresh huckleberries
- ½ c. yellow raisins
- ½ c. coarsely broken walnuts
- ½ tsp. ground cinnamon
- 1 c. fresh peach chunks
- 1 c. cantaloupe chunks
- 1 c. diced fresh green apple
- 1 lemon, seeded, cut into paper-thin slices
  sugar as directed

Directions:

1. In medium saucepan combine water and huckleberries.
2. Cover and cook over low heat, crushing berries occasionally, until tender.
3. Measure berries; for every cup, add ½ cup of sugar, stirring to melt.
4. Add walnuts and cinnamon; cover and remove from heat.
5. In large bowl combine peaches, cantaloupe, and apples.
6. Toss with lemon and let sit for 10 minutes, tossing a few times.
7. Pour huckleberry mixture over fruits.
8. Toss and serve for a delicious side dish with your meal.

Yields: 4 to 6 servings.

Did You Know? . . . .

Did you know that the Nez Perce boiled their dried huckleberries before they ate them?

Huckleberry Fresh Fruit Cocktail

Adding huckleberries gives a new twist to fruit cocktail. Serve this fruit cocktail as a special chilled side dish or as a sweet treat.

Ingredients:

2	c. apple juice
1	Tbs. lemon juice
½	tsp. lemon or orange zest
2	cinnamon sticks (3-in. lengths)
1	c. fresh huckleberries
1	c. fresh pineapple (may use canned if necessary)
2	Red Delicious apples, unpeeled, cored, cubed
1	orange, peeled, sectioned, halved
½	c. sm. seedless grapes or halve large ones
1	c. sour cream
¼	c. grated coconut
¼	c. apricot preserves
2	Tbs. dry white wine
½	c. macadamia nuts, chopped

Directions:

1. In medium saucepan combine apple juice, lemon juice, orange or lemon zest, and cinnamon sticks.
2. Heat to boiling, and simmer uncovered for 10 minutes.
3. Cool to room temperature.
4. In large serving bowl combine huckleberries, pineapple, apples, orange, and grapes.
5. Remove cinnamon sticks and pour apple juice mixture over fruit; let marinate for 30 minutes.
6. Combine sour cream, coconut, preserves, white wine, and nuts; blend well.
7. Completely drain fruits until no dripping is seen; combine with sour cream mixture.
8. Toss and serve immediately or chill until ready to serve.

Yields: 6 servings.

Huckleberry Onion Wild Rice Stuffing

This is a wonderful stuffing for any fowl with its delicious flavors mingled together.

Ingredients:

1	lb. breakfast sausage
½	onion, diced
½	green pepper, diced
3	c. bread cubes
½	c. chicken broth
1	c. cooked wild rice
½	c. dried huckleberries (recipe page 238)
4	slices bacon

Directions:

1. Preheat oven to 350 degrees F.
2. In skillet brown sausage, onions, and green pepper; drain off fat.
3. In large bowl toss together bread cubes and chicken broth until moist.
4. Add sausage, rice, and huckleberries, mixing thoroughly.
5. Place in greased casserole dish; lay strips of bacon over top and cover.
6. Bake for 20 minutes; remove cover and bake an additional 5 minutes or until bacon is browned and starts to crisp.
7. Remove from oven and serve while hot.

Yields: 4 servings.

Did You Know? . . . .

Did you know that when the Native American peoples dried their berries slowly over a fire that they kept smoldering in a rotten log, the bulk of the vitamin C content in the berries was preserved?

Multigrain Huckleberry Pilaf

We love wild rice, and the addition of huckleberries to this pilaf makes it very tasty and colorful.

Ingredients:

¾ c. pecans, chopped, toasted
⅔ c. wild rice
½ c. wheat berries
1 c. chopped onion
3 cloves garlic, finely chopped
2 Tbs. butter
1 Tbs. olive oil
2½ c. chicken broth
2½ tsp. rubbed sage
¼ tsp. ground pepper
⅓ c. brown rice
1½ c. huckleberries

Directions:

1. Preheat oven to 350 degrees F.
2. Spread pecans in ungreased pan; bake just until browned; set aside.
3. Rinse wild rice and wheat berries under cold running water; drain well.
4. In large saucepan cook onion and garlic in olive oil and butter over medium heat for about 10 minutes or until tender.
5. Stir in drained wild rice and wheat berries, chicken broth, sage, and pepper.
6. Bring to boiling then reduce heat; cover and simmer for 30 minutes.
7. Stir brown rice into wild rice mixture; return to boiling.
8. Reduce heat; cover and simmer for about 45 minutes or until grains are tender.
9. Stir huckleberries and pecans into rice mixture; serve warm.

Yields: 6 to 8 servings.

Huckleberry Raisin Dressing

This is delicious dressing for any meats or just by itself for any occasion or meal.

Ingredients:

- ½ c. chopped onion
- ½ c. chopped celery
- 3 Tbs. butter
- 1 c. huckleberries, fresh or frozen
- ½ c. golden raisins
- 4½ c. dry bread crumbs
- ½ tsp. sage
- ½ c. apple juice
-    ground black pepper to taste
-    vegetable oil

Directions:

1. Preheat oven to 350 degrees F.
2. Place onion, celery, and butter in skillet over medium-high heat.
3. Sauté for 5 minutes or until vegetables are tender.
4. Transfer into large bowl; add huckleberries, raisins, bread crumbs, sage, and pepper to taste, stirring well.
5. Add apple juice and toss lightly.
6. If dressing seems too dry, add more apple juice.
7. Turn into large, lightly oiled baking dish; cover and bake for 45 minutes.
8. Remove from oven and serve while hot.

Yields: 6 servings.

Did You Know? . . . .

Did you know that the Assiniboin people sometimes mixed huckleberries in pemmican?

Huckleberry Rice Pilaf

Huckleberries add color and flavor to this rice pilaf. It makes a great side dish for your favorite entrée.

Ingredients:

¾ c. chopped onion
1¼ c. chopped celery
½ c. huckleberries
⅔ c. chopped walnuts
1   Tbs. fresh thyme, chopped, or 1 tsp. dried
1   Tbs. fresh marjoram, chopped, or 1 tsp. dried
½ tsp. ground black pepper
1   Tbs. butter
3   c. cooked rice

Directions:

1. Put onion, celery, and huckleberries in nonstick skillet.
2. Add walnuts, thyme, marjoram, pepper, and butter.
3. Cook, uncovered, over medium heat for 10 minutes or until vegetables are tender; stirring occasionally.
4. Add rice; mix well.
5. Cook 3 to 4 minutes or until thoroughly heated.
6. Serve while hot with your favorite entrée.

Yields: 4 to 6 servings.

Did You Know? . . . .

Did you know that currently the House of the Montana legislature is working to define just what a wild Montana huckleberry is? A bill was introduced to bar the labeling of any product with "Montana" or "huckleberry" unless the berries used in the product are actually picked in Montana and are one of two species that are unique to the northwestern United States.

Huckleberry Dumplings

These are some great dumplings to serve with stewed chicken, turkey, or boiled pork roast. See how your family loves them!

Ingredients:

- 4 c. huckleberries
- 1½ c. water
- 2 c. sugar, divided
- 2 Tbs. butter, softened
- ½ tsp. vanilla extract
- ½ c. milk
- 1½ c. all-purpose flour
- 2 tsp. baking powder

Directions:

1. In large serving skillet over medium heat, bring berries, water, and 1½ cups sugar to a boil; reduce heat to simmer.
2. In large bowl cream remaining ½ cup sugar and butter together; add vanilla and milk, blending well.
3. Sift flour and baking powder together; gradually stir into milk mixture just until moistened.
4. Drop into simmering huckleberry mixture by tablespoonfuls; cover and simmer 20 minutes or until dumplings test done.
5. Remove serving skillet from stove, and place on table to serve.

Yields: 6 servings.

Did You Know? . . . .

Did you know that because huckleberries have not been able to be cultivated they can be considered a true "natural food" since no fertilizers or pesticides are used?

Pasta with Fresh Huckleberries, Tomatoes, and Corn

This is a wonderful summertime dish with the great taste of all those huckleberries, garden tomatoes, and fresh corn. It is really fast to prepare and tastes delicious.

Ingredients:

- 8 oz. pasta
- 5 Tbs. olive oil, divided
- 2 Tbs. red wine vinegar
- 1 tsp. dried basil
- 1 c. fresh huckleberries
- ½ c. whole corn kernels, cooked
- 2 tomatoes, chopped
- ½ c. chopped green onions or scallions
- 1 Tbs. grated Parmesan cheese
- 2 tsp. chopped fresh basil for garnish
  salt and ground black pepper to taste

Directions:

1. Cook pasta in large pot of boiling water until al dente, and drain completely; toss with 1 tablespoon olive oil then set aside.
2. Meanwhile, in large bowl whisk together remaining olive oil, red wine vinegar, and dried basil; add salt and pepper to taste.
3. Stir in huckleberries, corn, tomatoes, and scallions; let sit for 10 minutes to marinate flavors.
4. To serve, toss with pasta and sprinkle with grated Parmesan cheese and fresh basil, if desired.

Yields: 4 servings.

Wild Rice with Huckleberries and Mushrooms

This is such an easy and delicious dish to make while you go about preparing the rest of the meal.

Ingredients:

2   c. chicken broth
1   c. water
1   c. wild rice
1   c. huckleberries, fresh or frozen
½   c. seedless raisins
1   c. fresh mushrooms
    additional huckleberries for garnish

Directions:

1. Bring water and chicken broth to a boil in saucepan with lid.
2. Add rice and cook until al dente tender.
3. Stir in huckleberries, raisins, and mushrooms.
4. Turn heat to simmer or low, and cook until tender, stirring only once to assure that mixture is not burning on bottom.
5. Cook for 25 minutes; turn heat off and leave covered until ready to serve.
6. When ready to serve, sprinkle some fresh huckleberries over top.

Yields: 6 servings.

Did You Know? . . . .

Did you know that for the most part huckleberries are hand-picked in their native habitat, although in some areas experimentation has been done with mechanical harvesters?

Rosemary Roasted Sweet Potatoes
with Huckleberries

Huckleberry and rosemary add a sweet flavor to this dish.

Ingredients:

- 3 Tbs. extra-virgin olive oil
- 2 tsp. kosher salt
- ¼ tsp. black pepper
- 2 tsp. very finely minced fresh rosemary
- 4 lg. sweet potatoes (about 4½ lb.)
- 1 med. white onion
- ½ c. wild huckleberries
  fresh rosemary sprigs for garnish

Directions:

1. Preheat oven to 400 degrees F.
2. Stir olive oil, salt, pepper, and minced rosemary together in large bowl.
3. Peel sweet potatoes, cut in half lengthwise, then cut each half into 6 large, chunky pieces; add to bowl and set aside.
4. Peel onion and trim root end but keep it intact.
5. Cut onion in half vertically, then cut each half into about 6 wedges; add to bowl.
6. With rubber spatula mix potatoes and onions with olive oil mixture to coat vegetables well.
7. Transfer mixture to rimmed baking sheet, and bake for 40 minutes.
8. Remove from oven, and with spatula turn potatoes and onions over.
9. Increase oven to 450 degrees F.; return pan to oven and bake for another 15 minutes or until potatoes are crisp and golden.
10. Sprinkle huckleberries over potatoes and return to oven for 5 minutes more.
11. Transfer to warmed platter and garnish with sprigs of fresh rosemary.

Yields: 6 servings.

Steamed Huckleberry Pudding

This is a great dish for any meal and makes a delicious side dish for a holiday or special occasion dinner.

Ingredients:

- 4 c. all-purpose flour
- ½ tsp. salt
- 4 tsp. baking powder
- 2 Tbs. butter
- 1 c. milk
- 2 c. huckleberries
- 1 c. golden raisins
- ½ c. chopped sweet onion
- 2 Tbs. brandy of choice

Directions:

1. Preheat oven to 375 degrees F.
2. Sift together flour, salt, and baking powder; cut in butter.
3. Add milk, berries, raisins, onion, and brandy, stirring well.
4. Spoon into large, deep, buttered casserole dish with lid.
5. Place casserole dish in deep pan of water to cover ⅔ of the way up side of dish.
6. Bake for 1½ hours or until knife inserted in center comes out clean.
7. Remove from oven and serve while hot.

Yields: 8 servings.

Did You Know? . . . .

Did you know that the box huckleberry grows in a low, dark-green carpet, has pinkish flowers, light-blue fruit, and can spread up to 6 inches per year by underground runners?

Huckleberry Delights Cookbook

A Collection of Huckleberry Recipes
Cookbook Delights Series Book 6

Soups

Table of Contents

Chilled Cinnamon Huckleberry Soup

This makes a very pretty, cold soup on a hot summer day. Select your favorite garnishment of whipped or sour cream, or try a little of both.

Ingredients:

- 6 c. fresh huckleberries, washed, hulled
- 3 c. water
- ⅓ c. sugar
- 1 cinnamon stick (2-in. length)
- 1 lemon slice
- 1 Tbs. cornstarch dissolved in 2 tsp. water
- 1 c. heavy whipping cream
- ½ c. fruity, semidry white wine or Catawba juice
  whipped or sour cream

Directions:

1. In 3-quart saucepan combine berries, water, sugar, cinnamon stick, and lemon slice.
2. Boil slowly until berries are soft and have rendered their color.
3. Stir in dissolved cornstarch; cook, stirring constantly, until thick.
4. Discard cinnamon stick and lemon slice.
5. Transfer soup to blender or food processor; purée mixture.
6. Return soup to saucepan; stir in heavy cream and wine.
7. Cover and chill thoroughly.
8. Serve in chilled soup bowls or mugs, and garnish with a spoonful of whipped or sour cream.

Yields: 10 to 12 servings.

Did You Know? . . . .

Did you know that the foliage of the evergreen huckleberry plant is often used in flower arrangements?

Huckleberry Dumpling Soup

This is a nice soup for a leisurely weekend brunch or a great beginning for lunch or dinner when the weather is exceptionally warm. It is also good for a quick warm up when served hot during cold weather.

Ingredients for soup:

4	c. huckleberries
⅓	c. sugar
1	c. water
2	tsp. lemon juice

Ingredients for dumplings:

1	c. all-purpose flour
2	tsp. baking powder
3	tsp. sugar
2	Tbs. butter
½	c. milk

Directions for soup:

1. Simmer huckleberries, sugar, and water for 10 minutes.
2. Add lemon juice and adjust heat to simmer.

Directions for dumplings:

1. Sift together flour, baking powder, and sugar.
2. Cut in butter until well mixed; add milk and mix just until moistened.
3. Drop batter by spoonfuls into simmering berry mixture.
4. Cover tightly; cook 15 minutes.
5. Remove from heat and serve hot, or refrigerate for cold soup.

Yields: 4 to 6 servings.

Fresh Huckleberry Soup

This cold huckleberry soup uses orange juice and honey instead of liqueur and sugar. It is great for children as well as adults.

Ingredients:

1   c. yogurt, plain or vanilla
1   c. orange juice
1½ Tbs. honey
5   c. fresh huckleberries

Directions:

1. In small mixing bowl whisk all ingredients until blended.
2. Cover and refrigerate until chilled.
3. Pour into individual bowls.

Huckleberry Soup with Peaches

This is a tasty summer soup to make when both huckleberries and peaches are in season. Make it fresh no more than two hours before serving.

Ingredients:

4   c. huckleberries, divided
½   c. water
6   Tbs. sugar, divided
1   tsp. kirsch or cherry brandy
2   c. fresh peaches, peeled, sliced

Directions:

1. In medium, nonreactive saucepan, combine 3 cups huckleberries and water.
2. Bring to a simmer, cover, and cook 15 minutes or until huckleberries are very soft.
3. Press huckleberry mixture through sieve or through doubled cheesecloth over small bowl; reserve liquid.

4. Combine huckleberry liquid, 3 tablespoons sugar, and kirsch.
5. Cover; chill mixture at least 2 hours or until ready to serve.
6. Combine remaining sugar and peaches; toss gently to coat.
7. Spoon ¼ cup huckleberry mixture into each of 6 shallow fruit or soup bowls.
8. Arrange ⅓ cup peach slices and remaining huckleberries over each serving.

Yields: 6 servings.

Huckleberry Wine Soup

This soup has combined the flavor of burgundy with the delicious taste of huckleberries.

Ingredients:

4	c. huckleberries
2	c. water
1	c. burgundy wine
¾	c. sugar
¼	tsp. ground cinnamon
1	c. sour cream, divided
	fresh mint

Directions:

1. In large saucepan combine huckleberries, water, wine, sugar, and cinnamon.
2. Bring to a boil over moderate heat, stirring occasionally.
3. Reduce heat, cover, and simmer 3 minutes longer.
4. Allow to cool, then purée in blender until smooth.
5. Add ¾ cup sour cream and blend well.
6. Pour into serving bowls and chill.
7. Just before serving, spoon remaining sour cream on top and garnish with fresh mint.

Yields: 6 servings.

Honeydew Huckleberry Soup

You and your family or guests are bound to appreciate this simple and delicious cold soup of honeydew and huckleberries.

Ingredients:

- 1   honeydew melon, peeled, cut into chunks
- 1   pt. huckleberries
- 6   oatmeal cookies
  Sweetened Whipped Cream (recipe page 167)

Directions:

1. Purée melon until smooth in food processor or blender.
2. Pour into large bowl, and stir huckleberries into puréed melon.
3. Cover and chill in refrigerator until quite cold.
4. To serve, ladle soup into individual bowls and crumble an oatmeal cookie over each serving or top with whipped cream and serve cookies alongside.

Yields: 6 servings.

Easy Creamy Huckleberry Soup

This is an easy-to-make, colorful soup that is a delightful way to start a meal on a hot summer day.

Ingredients:

- 2   pt. huckleberries
- 1⅔ c. water
- ½   c. honey
- 1   c. plain yogurt
  pinch of ground cinnamon
  additional huckleberries for garnish

Directions:

1. In medium saucepan combine huckleberries, water, and honey.
2. Heat to gentle simmer, and cook until fruit breaks down and softens, about 20 minutes.
3. Cool to lukewarm, then purée mixture in food processor or blender.
4. Transfer to mixing bowl and stir in yogurt; season with cinnamon.
5. Cover and refrigerate until chilled through, at least 3 hours or until ready to serve.

Yields: 4 servings.

Cold Huckleberry Soup

Here is another version of a chilled soup. This one has the added flavor of Amaretto.

Ingredients:

6   c. huckleberries
1   c. heavy cream
¼   tsp. ground cinnamon
1   tsp. lemon juice
¾   oz. Amaretto liqueur
1   lemon, slice half, juice other half

Directions:

1. In blender combine huckleberries, cream, cinnamon, and lemon juice.
2. Blend until smooth then stir in liqueur.
3. Cover and place in refrigerator; chill completely until ready to serve.
4. Serve chilled, garnished with lemon slices.

Yields: 4 servings.

Huckleberry Soup with Crème Fraîche

This version of cold huckleberry soup uses crème fraîche, which sounds difficult but is very easy to make.

Ingredients:

- 8 c. fresh huckleberries
- 2 c. orange juice
- 1 c. crème fraîche (made with half whipping cream and half sour cream)
  sugar to taste
  huckleberries for garnish

Directions:

1. Place huckleberries in food processor; purée until smooth.
2. Pour into bowl; add orange juice and sugar, blending well.
3. Whisk in crème fraîche until well blended.
4. Cover and chill in refrigerator for 2 hours or until ready to serve.
5. Serve in chilled bowls, garnished with fresh huckleberries.

Yields: 6 to 8 servings.

Huckleberry Champagne Soup

Here is a summertime cold soup to beat the heat and warm the senses. None could be simpler to make than this one.

Ingredients:

- 5 c. fresh huckleberries
- ¼ c. sugar
- 1 c. champagne
  Sweetened Whipped Cream (recipe page 167)

Directions:

1. Place huckleberries into blender and sprinkle with sugar; process until smooth.
2. Cover and chill for 2 hours or until ready to serve.
3. Stir in champagne just before serving.
4. Pour into serving bowls, and add dollop of sweetened whipped cream if desired.

Yields: 4 servings.

Huckleberry Burgundy Soup

You will want to use those hard-earned huckleberries that taste so good when you have picked them yourself. You will enjoy using them in this delicious soup, which can be served hot or cold.

Ingredients:

4   c. fresh huckleberries
1   c. sugar
1   c. sour cream
4   c. cold water
1   c. burgundy or other dry red wine
    nutmeg for garnish

Directions:

1. Use food processor or blender to purée huckleberries.
2. Pour purée into large saucepan; stir in sugar, sour cream, and water.
3. Cook over medium-low heat, stirring gently, for 20 to 25 minutes to fully blend flavors; do not allow boiling.
4. Stir in burgundy; serve warm or chilled with a shake of nutmeg over the top.

Yields: 8 servings.

Huckleberry Barley Soup

This is one of those satisfying soups to linger over with the family, especially on a warm summer evening out on the deck.

Ingredients:

- ½ c. barley
- 6 c. water
- ½ c. sugar
- 2 c. huckleberries, fresh or frozen
- ½ c. raisins
- 1 c. pitted sweet cherries
  additional huckleberries for garnish
  ground cinnamon for garnish

Directions:

1. In large bowl soak barley in water overnight; do not drain.
2. Place in large saucepan over low heat; simmer for 1 hour.
3. Add sugar, huckleberries, and raisins; simmer for another 30 minutes.
4. Add cherries and continue to simmer for another 15 minutes or until soup becomes relatively thick.
5. Remove from heat; cool, then allow to chill in refrigerator until ready to serve.
6. Can be stored, covered, in refrigerator for up to 1 week.
7. If desired, sprinkle additional fresh huckleberries over the top along with a shake of cinnamon.

Yields: 4 to 6 servings.

Did You Know? . . . .

Did you know that when the black hawthorn fruits were ripe, the Okanagan-Colville people knew that meant that the black huckleberries would be ripe in the mountains?

Huckleberry Delights Cookbook

A Collection of Huckleberry Recipes
Cookbook Delights Series Book 6

Wines and Spirits

Table of Contents

About Cooking with Alcohol

Some recipes in this cookbook contain, among other ingredients, liquors. It is for the purpose of obtaining desired flavor and achieving culinary appreciation and not to be abused in any way. In cooking and baking, alcohol evaporates and only the flavor may be enjoyed. When mixed in cold, however, such as in desserts, caution must be exercised. These recipes are intended for people who may consume small amounts of alcohol in a responsible and safe manner.

I live in Washington State and we are proud of our wine production. Washington State is rapidly gaining prestige as a premier wine producer. Do enjoy the art of wine tasting and enjoy the completeness and uniqueness of each wine. It is an art to enjoy and savor in moderation.

If consumption of even small amounts of alcoholic ingredients presents a problem, in whatever form, please substitute coffee flavor syrups, found in coffee sections of supermarkets. For example, instead of Southern Comfort liqueur, substitute with Irish Cream or Amaretto Syrup.

Karen Jean Matsko Hood

Huckleberry Kir

This is a simple yet tasteful drink to serve to your guests during or after an evening meal.

Ingredients:

2⅔ c. Chablis or other white wine, chilled
1   Tbs. huckleberry liqueur (recipe page 302)

Directions:

1. Pour ⅔ cup wine into each chilled wine glass.
2. Add ¾ teaspoon liqueur to each one, and stir well.

Yields: 4 servings.

Huckleberry Wine

It can be an entertaining project to make your own homemade wine and watch the fermentation process. Make sure all the items are totally clean and sanitized. You may need to order yeast and Campden tablets from a mail order catalog or website if you cannot find them in a local store. This will be a very smooth wine.

Ingredients:

4   lb. huckleberries
2½   lb. sugar
1½   tsp. acid blend
1   tsp. yeast nutrient
1   Campden tablet, crushed
7¼   pt. water
1   pkg. champagne wine yeast

Directions:

1. Put water on to boil.
2. Meanwhile, sort and wash berries, discarding any not sound or ripe.
3. Put huckleberries in primary, and mash with sanitized potato masher or piece of hardwood.
4. Add sugar to primary, and pour boiling water over berries and sugar, stirring to dissolve.
5. Cover with sanitized cloth, and set aside to cool to room temperature.
6. When cool, add remaining ingredients except yeast.
7. Stir, cover primary, and set aside 24 hours.
8. Add activated yeast.
9. When fermentation is vigorous, stir twice daily for 10 days.
10. Strain through nylon straining bag without squeezing.
11. Drip drain 30 to 45 minutes and pour juice into secondary.
12. Attach airlock and set aside.
13. Rack every 60 days for 6 months, topping up and refitting airlock each time.
14. At last racking, rack into bottles or stabilize, sweeten to taste, wait 10 days, and rack into bottles.

Huckleberry Lift

This drink is light and refreshing on a breezy summer evening.

Ingredients:

  3   oz. sparkling wine, chilled
  ½   oz. grenadine syrup
  6   whole huckleberries
      bottled water

Directions:

1. Fill champagne flute half full with chilled sparkling wine.
2. Add grenadine then fill glass with bottled water.
3. Drop in huckleberries.

Yields: 1 serving.

Huckleberries with Orange Liqueur and Lavender

Lavender flower heads add an old-fashioned taste and look to this huckleberry liqueur.

Ingredients:

  1   c. orange-flavored liqueur
  1   c. water
  1   c. sugar
  1½   lb. fresh huckleberries
  20   fresh lavender flower heads

Directions:

1. Prepare jars, lids, and boiling water bath.
2. Combine liqueur, water, and sugar in pan.
3. Cook over medium-high heat, stirring frequently until sugar is dissolved and mixture has time to boil.

4. Remove from heat.
5. Pick over, wash, and dry huckleberries, then pack in hot, dry jars, placing 4 lavender heads in each jar; leave ½-inch headspace.
6. Pour hot liquid into jars, just covering berries.
7. Wipe rims with clean towel and attach lids securely.
8. Place jars in boiling water bath and, when water returns to a full boil, process for 15 minutes.
9. Let stand for 1 week before using.

Yields: 5 half-pints.

Frozen Huckleberry Margarita

This is one of the most refreshing drinks you can enjoy on a hot summer night!

Ingredients:

2	tsp. coarse salt
1	lime wedge
3	oz. white tequila
1	oz. triple sec
2	oz. lime juice
1	oz. huckleberry juice (recipe page 242)
1	c. crushed ice

Directions:

1. Place salt in saucer.
2. Rub rim of cocktail glass with lime wedge, and dip glass into salt to coat rim thoroughly.
3. Reserve lime.
4. Pour tequila, triple sec, lime juice, huckleberry juice, and crushed ice into blender; blend well at high speed.
5. Pour into the cocktail glass and serve.

Yields: 1 serving.

Huckleberry Fuzzy Navel

Many of us have tried a fuzzy navel, but have you tried one with a huckleberry flavor?

Ingredients:

> 1½ oz. peach schnapps
> orange juice to taste
> huckleberry juice to taste
> ice

Directions:

1. Pour peach schnapps into ice-filled Collins glass.
2. Fill with huckleberry juice and orange juice, stirring to combine.
3. Adjust amounts of juices to taste.

Yields: 1 serving.

Huckleberry Eggnog

My family enjoys eggnog, and this makes a colorful drink to enjoy over the holidays or year round.

Ingredients:

> 3 c. whole milk
> 7 lg. eggs
> 1 c. sugar
> 2 c. heavy cream
> 1 tsp. vanilla extract
> ⅓ c. huckleberry juice (recipe page 242)
> ⅓ c. Cognac or other brandy (optional)
> freshly grated nutmeg

Directions:

1. Bring milk just to a boil in 2-quart heavy saucepan.
2. Whisk together eggs and sugar in large bowl.
3. Add hot milk in slow stream, whisking.
4. Pour mixture into saucepan and cook over moderately low heat, stirring constantly with wooden spoon, until mixture registers 170 degrees F. on thermometer, 6 to 7 minutes.
5. Pour custard through fine-mesh sieve into clean large bowl and stir in cream and vanilla; add huckleberry juice and brandy, if desired.
6. Cool completely, uncovered, then chill, covered, until cold, at least 3 hours and up to 24.
7. Serve sprinkled with freshly grated nutmeg.
8. Note: Flavor of eggnog improves when made a day ahead to allow alcohol to mellow.

Yields: About 6 cups.

Huckleberry Margaritas

This is a creative twist on the classic margarita, and the huckleberries add an attractive color.

Ingredients:

1½ oz. tequila
½   oz. triple sec
1   oz. lime juice
1   oz. huckleberry juice (recipe page 242)
    salt
    ice

Directions:

1. Rub rim of cocktail glass with lime juice, then dip rim in salt.
2. Place tequila, triple sec, lime juice, and huckleberry juice in shaker jar with ice.
3. Shake well, strain into salt-rimmed glass, and serve.

Yields: 1 serving.

Huckleberry Twist

This drink has a strong, bold flavor and is good at any time of the year.

Ingredients:

- 1 oz. peppermint schnapps
- 1 oz. blackberry brandy
- 2-3 oz. huckleberries
- 1 pint ice
- 1 lemon slice

Directions:

1. Blend all ingredients except lemon slice together until smooth.
2. Pour into highball glass; garnish with lemon slice.

Yields: 1 serving.

Huckleberry Vodka Martini

These huckleberry martinis will be the hit of your party or gathering with a splash of lime juice to give it zip! The key is the homemade huckleberry vodka that requires 2 weeks to properly infuse.

Ingredients:

- 1 liter vodka
- 1 pint huckleberries, rinsed, dried
- 1 c. raspberry-flavored liqueur
- 1 lime, juiced
- 1 twist lime zest, garnish

Directions:

1. To make huckleberry vodka, pour out approximately ⅓ bottle of vodka into holding container; set aside.

2. Score each huckleberry with a small nick, and place into vodka bottle.
3. With the vodka previously set aside, fill vodka bottle until just below neck.
4. Add just enough raspberry liqueur to top off bottle; let sit in dark place for 2 weeks.
5. To make martinis, in cocktail shaker filled with ice, combine 2 parts huckleberry vodka, 1 part raspberry liqueur, and a dash of lime juice.
6. Shake vigorously and strain into glass.
7. Garnish with twist of lime zest.

Yields: 10 servings.

Huckleberry Singapore Sling

This is a very fruity concoction of a very familiar drink with the added flavor and color of huckleberries.

Ingredients:

- ½ oz. grenadine
- 1 oz. gin
- 2 oz. sweet and sour
- 2 oz. huckleberry juice (recipe page 242)
- ½ oz. cherry brandy
- 1 cherry
  carbonated water
  ice cubes

Directions:

1. Pour grenadine, gin, sweet and sour, and huckleberry juice over ice cubes and stir well.
2. Fill with carbonated water and top off with cherry brandy.
3. Add cherry on top and serve.

Yields: 1 serving.

Huckleberry Black Russian

Huckleberry flavor actually blends quite well with the classic Black Russian. Try this unique version.

Ingredients:

- ¾ oz. coffee liqueur
- 1½ oz. vodka
- 1½ oz. huckleberry juice (recipe page 242)
  ice cubes

Directions:

1. Pour liqueur, vodka, and huckleberry juice over ice cubes in an old-fashioned glass and serve.

Yields: 1 serving.

Huckleberry Daiquiri

This makes a refreshing version of the daiquiri drink and will be a hit with your guests!

Ingredients:

- 1¼ oz. white rum
- ¾ c. huckleberries
- 1 Tbs. simple syrup, made of ½ water and ½ sugar
- ¼ c. orange juice
- 1½ c. ice
  orange slice and cherry for garnish

Directions:

1. Blend rum, huckleberries, syrup, and orange juice in blender until mixed.

2. Add ice a bit at a time and blend until smooth.
3. Pour into tall, chilled glass and garnish with an orange slice and cherry.

Yields: 1 serving.

Huckleberry Sangria

This is a flavorful variation on the classic sangria. Try this as an easy way to make your dinner more festive.

Ingredients:

2	oranges
2	lemons
2	bottles dry white wine
4	Tbs. sugar
2	oz. brandy
2	oz. Cointreau
1½	c. huckleberry juice (recipe page 242)
4	c. ice cubes
2	c. club soda

Directions:

1. Cut oranges in half, cut one half into thin slices and juice remaining halves.
2. Cut lemons into thin slices and combine in large pitcher with orange juice, wine, sugar, brandy, Cointreau, and huckleberry juice.
3. Chill until ready to serve.
4. To serve, add ice and club soda, stir gently, and pour into chilled glasses.

Yields: 6 to 8 servings.

Huckleberry Liqueur

You will enjoy the outstanding flavor of the wild huckleberries in this drink.

Ingredients:

- 1   bottle high-quality vodka (750 ml)
- 1½ c. huckleberries, washed
- 1   lemon, washed
- 1½ c. sugar
- 1   c. water

Directions:

1. Divide vodka and washed berries between 2 clean glass quart jars with lids.
2. With potato peeler, peel zest from lemon, being sure not to get any white pith.
3. Divide zest between jars. (Use remainder of lemon for another purpose.)
4. Cap jars and shake well.
5. Let sit at room temperature for 2 weeks, shaking jars every couple of days.
6. After 2 weeks, bring sugar and water to a boil in large saucepan, stirring to dissolve sugar.
7. Boil for 2 minutes, then let cool to room temperature.
8. Using fine-mesh strainer, strain vodka from huckleberries and zest into large bowl to catch every drop. (Be sure to press out all the liquid you can from berries before you discard them.)
9. Mix cooled sugar syrup into strained liquor; stir to combine.
10. Bottle liqueur in fancy bottles or in clean, clear wine bottles.
11. Cork and store for up to one year.
12. Note: If you like your liqueur a bit sweeter, increase the sugar to 2 cups.

Yields: 5 to 6 cups.

Huckleberry Schnapps

You can make your own huckleberry schnapps to have on hand to use in a variety of drinks. For best results use fully ripe, fresh huckleberries.

Ingredients:

  huckleberries
  unflavored vodka

Directions:

1. Rinse berries carefully; leave to dry on paper towel.
2. Using a clean glass jar with a tight-fitting lid, fill ½ to ⅔ full with huckleberries.
3. Fill jar with clear, unflavored vodka – 40% alcohol (80 proof).
4. Steep at least 3 months in a dark place at room temperature (64 to 68 degrees F.).
5. Shake lightly and taste from time to time.
6. Strain and filter into clean glass bottle or jar with tight-fitting lid; allow to settle for 2 days.
7. Schnapps may also be stored (aged) 4 to 6 months or longer in a dark place at room temperature. (Flavor changes during storage. Short storage schnapps has more of a fruity taste.)
8. Filter out sediments from time to time as needed.
9. Note: If not satisfied with your infusion you may adjust as follows.
10. If flavor is too strong, dilute with vodka; allow to settle 2 days or more before tasting and making further adjustments if necessary.
11. If flavor is too weak, enhance by adding very small amounts of a simple sugar syrup.
12. Allow to settle 2 days then taste, adding more sugar syrup as needed. (Never use artificial sweeteners.)
13. Honey may also be used as a sweetener, but make sure the honey flavor goes well with huckleberry.
14. Use 1 tablespoon honey per liter of infusion; melt in small saucepan and add little by little, tasting after each addition.
15. Shake (do not whisk or blend), and allow to settle as before.

Festival Information

Following is a list of just some of the huckleberry festivals throughout the country each year. You may use the following contact information or contact the local Chamber of Commerce or Visitor's Information Bureau of each town to find out the exact dates for the festival for that community.

Huckleberry Days
August each year
Trout Creek, MT
Phone: 406-827-3301 or 406-827-5077
Website: http://huckleberryfestival.com

Huckleberry Festival
September each year
Bingen, WA
Phone: 509-493-5294
Email: huckleberryfest@yahoo.com

Mount Hood Huckleberry Festival
August each year
Welches, OR
Phone: 503-622-4798
Email: cgsmthood@onemain.com

Shawangunk Mountain Wild Blueberry and Huckleberry Festival
August each year
Ellenville, NY
Phone: 845-647-4620
Email: chamberofcommerce2@hvc.rr.com

Wallace Heritage-Huckleberry Festival
August each year
Wallace, ID
Phone: 208-753-7151 or 800-434-4204
Website: www.wallaceidahochamber.com

U.S. and Metric Measurement Charts

Here are some measurement equivalents to help you with exchanges. There was a time when many people thought the entire world would convert to the metric scale. While most of the world has, America still has not. Metric conversions in cooking are vitally important to preparing a tasty recipe. Here are simple conversion tables that should come in handy.

U.S. Measurement Equivalents

A few grains/pinch/dash, (dry) = Less than ⅛ teaspoon

A dash (liquid) = A few drops

3 teaspoons = 1 tablespoon

½ tablespoon = 1½ teaspoons

1 tablespoon = 3 teaspoons

2 tablespoons = 1 fluid ounce

4 tablespoons = ¼ cup

5⅓ tablespoons = ⅓ cup

8 tablespoons = ½ cup

8 tablespoons = 4 fluid ounces

10⅔ tablespoons = ⅔ cup

12 tablespoons = ¾ cup

16 tablespoons = 1 cup

16 tablespoons = 8 fluid ounces

⅛ cup = 2 tablespoons

¼ cup = 4 tablespoons

¼ cup = 2 fluid ounces

⅓ cup = 5 tablespoons plus 1 teaspoon

½ cup = 8 tablespoons

1 cup = 16 tablespoons

1 cup = 8 fluid ounces

1 cup = ½ pint

2 cups = 1 pint

2 pints = 1 quart

4 quarts (liquid) = 1 gallon

8 quarts (dry) = 1 peck

4 pecks (dry) = 1 bushel

1 kilogram = approximately 2 pounds

1 liter = approximately 4 cups or 1 quart

Approximate Metric Equivalents by Volume

U.S.	Metric
¼ cup	= 60 milliliters
½ cup	= 120 milliliters
1 cup	= 230 milliliters
1¼ cups	= 300 milliliters
1½ cups	= 360 milliliters
2 cups	= 460 milliliters
2½ cups	= 600 milliliters
3 cups	= 700 milliliters
4 cups (1 quart)	= .95 liter
1.06 quarts	= 1 liter
4 quarts (1 gallon)	= 3.8 liters

Approximate Metric Equivalents by Weight

U.S.	Metric
¼ ounce	= 7 grams
½ ounce	= 14 grams
1 ounce	= 28 grams
1¼ ounces	= 35 grams
1½ ounces	= 40 grams
2½ ounces	= 70 grams
4 ounces	= 112 grams
5 ounces	= 140 grams
8 ounces	= 228 grams
10 ounces	= 280 grams
15 ounces	= 425 grams
16 ounces (1 pound)	= 454 grams

Glossary

Aerate: A synonym for sift; to pass ingredients through a fine-mesh device to break up large pieces and incorporate air into ingredients to make them lighter.

Al dente: "To the tooth," in Italian. The pasta is cooked just enough to maintain a firm, chewy texture.

Amaretto: A liqueur with a distinct flavor of almonds, though it is often made with apricot pits kernels.

Baste: To brush or spoon liquid fat or juices over meat during roasting to add flavor and prevent drying out.

Bias-slice: To slice a food crosswise at a 45-degree angle.

Blanch: To scald, as in vegetables being prepared for freezing; as in almonds so as to remove skins.

Blend: To mix or fold two or more ingredients together to obtain equal distribution throughout the mixture.

Braise: To brown meat in oil or other fat and then cook slowly in liquid. The effect of braising is to tenderize the meat.

Bread: To coat food with crumbs (usually with soft or dry bread crumbs), sometimes seasoned.

Brown: To quickly sauté, broil, or grill either at the beginning or end of meal preparation, often to enhance flavor, texture, or eye appeal.

Caramelization: Browning sugar over a flame, with or without the addition of some water to aid the process. The temperature range in which sugar caramelizes is approximately 320 to 360 degrees F.

Chiffonade: To cut herbs or leafy green vegetables into long, thin strips, generally by stacking the leaves, rolling them tightly, then cutting across the rolled leaves with a sharp knife.

Clarify: To remove impurities from butter or stock by heating the liquid, then straining or skimming it.

Confit: To slowly cook pieces of meat in their own gently rendered fat.

Core: To remove the inedible center of fruits such as pineapples.

Cream: To beat vegetable shortening, butter, or margarine, with or without sugar, until light and fluffy. This process traps in air bubbles, later used to create height in cookies and cakes.

Crimp: To create a decorative edge on a piecrust. On a double piecrust, this also seals the edges together.

Curd: Custard-like pie or tart filling flavored with juice and zest of citrus fruit, usually lemon, although lime and orange may also be used.

Curdle: To cause semisolid pieces of coagulated protein to develop in food, usually as a result of the addition of an acid substance, or the overheating of milk or egg-based sauces.

Custard: A mixture of beaten egg, milk, and possibly other ingredients such as sweet or savory flavorings, which is cooked with gentle heat, often in a water bath or double boiler. As pie filling, the custard is frequently cooked and chilled before being layered into a baked crust.

Deglaze: To add liquid to a pan in which foods have been fried or roasted, in order to dissolve the caramelized juices stuck to the bottom of the pan.

Dot: To sprinkle food with small bits of an ingredient such as butter to allow for even melting.

Dredge: To sprinkle lightly and evenly with sugar or flour. A dredger has holes pierced on the lid to sprinkle evenly.

Drizzle: To pour a liquid such as a sweet glaze or melted butter in a slow, light trickle over food.

Drippings: The liquids left in the bottom of a roasting or frying pan after meat is cooked. Drippings are used for gravies and sauces.

Dust: To sprinkle food lightly with spices, sugar, or flour for a light coating.

Egg Wash: A mixture of beaten eggs (yolks, whites, or whole eggs) with either milk or water. Used to coat cookies and other baked goods to give them a shine when baked.

Emulsion: A mixture of liquids, one being a fat or oil and the other being water based so that tiny globules of one are suspended in the other. This may involve the use of stabilizers, such as egg or custard. Emulsions may be temporary or permanent.

Entrée: A French term that originally referred to the first course of a meal, served after the soup and before the meat courses. In the United States, it refers to the main dish of a meal.

Fillet: To remove the bones from meat or fish for cooking.

Filter: To remove lumps, excess liquid, or impurities by passing through paper or cheesecloth.

Firm-Ball Stage: In candy making, the point where boiling syrup dropped in cold water forms a ball that is compact yet gives slightly to the touch.

Flambé: To ignite a sauce or other liquid so that it flames.

Flan: An open pie filled with sweet or savory ingredients; also, a Spanish dessert of baked custard covered with caramel.

Flute: To create a decorative scalloped or undulating edge on a piecrust or other pastry.

Framboise: A brandy or liqueur made from raspberries.

Fricassee: Usually a stew in which the meat is cut up, lightly cooked in butter, and then simmered in liquid until done.

Frizzle: To cook thin slices of meat in hot oil until crisp and slightly curly.

Fromage Blanc: A dairy product originating from Belgium and the north of France. It literally means "white cheese." Fromage blanc is unlike cheese, however, in that the curds are not allowed to solidify, giving it a texture similar to yogurt. It is also known as fromage frais.

Ganache: A rich chocolate filling or coating made with chocolate, vegetable shortening, and possibly heavy cream. It can coat cakes or cookies and be used as a filling for truffles.

Glaze: A liquid that gives an item a shiny surface. Examples are fruit jams that have been heated or chocolate thinned with melted vegetable shortening. Also, to cover a food with such a liquid.

Gratin: To bind together or combine food with a liquid such as cream, milk, béchamel sauce, or tomato sauce, in a shallow dish. The mixture is then baked until cooked and set.

Hard-Ball Stage: In candy making, the point at which syrup has cooked long enough to form a solid ball in cold water.

Haupia: A traditional coconut milk-based Hawaiian dessert. Although it is technically a pudding, its consistency is more like gelatin desserts and can be served in blocks like gelatin.

Hull (also husk): To remove the leafy parts of soft fruits, such as strawberries or blackberries.

Infusion: To extract flavors by soaking them in liquid heated in a covered pan. The term also refers to the liquid resulting from this process.

Julienne: To cut into long, thin strips.

Jus: The natural juices released by roasting meats.

Kalamata Olive: A large, black olive, named after the city of Kalamata, Greece. It has a smooth, meat-like taste and is used as a table olive.

Larding: To inset strips of fat into pieces of meat, so that the braised meat stays moist and juicy.

Marble: To gently swirl one food into another.

Marinate: To combine food with aromatic ingredients to add flavor.

Meringue: Egg whites beaten until they are stiff, then sweetened. It can be used as the topping for pies or baked as cookies.

Mull: To slowly heat cider with spices and sugar.

Nonreactive Pan: Cookware that does not react chemically with foods, primarily acidic foods. Glass, stainless steel, enamel, anodized aluminum, and permanent nonstick surfaces are basically nonreactive. Shiny aluminum is reactive.

Panna Cotta: An Italian phrase meaning "cooked cream." It is a softly set, creamy Italian pudding. It is served with wild berries, caramel, or chocolate sauce.

Parboil: To partly cook in a boiling liquid.

Peaks: The mounds made in a mixture. For example, egg white that has been whipped to stiffness. Peaks are "stiff" if they stay upright or "soft" if they curl over.

Pesto: A sauce usually made of fresh basil, garlic, olive oil, pine nuts, and cheese. The ingredients are finely chopped and then mixed, uncooked, with pasta. Generally, the term refers to any uncooked sauce made of finely chopped herbs and nuts.

Pierogi: A filled Slavic semi-circular dumpling of unleavened dough. Mashed potatoes are the most common filling, but they may be stuffed with a variety of vegetables, meat, eggs, cheese, or fruit.

Pipe: To force a semisoft food through a bag (either a pastry bag or a plastic bag with one corner cut off) to decorate food.

Pressure Cooking: To cook using steam trapped under a locked lid to produce high temperatures and achieve fast cooking time.

Purée: To mash or sieve food into a thick liquid.

Ramekin: A small baking dish used for individual servings of sweet and savory dishes.

Reduce: To cook liquids down so that some of the water evaporates.

Refresh: To pour cold water over freshly cooked vegetables to prevent further cooking and to retain color.

Rolling Boil: A boil that does not stop bubbling when stirred.

Roux: A cooked paste usually made from flour and butter used to thicken sauces.

Sauté: To cook foods quickly in a small amount of oil in a skillet or sauté pan over direct heat.

Scald: To heat a liquid, usually a dairy product, until it almost boils.

Sear: To seal in a meat's juices by cooking it quickly using very high heat.

Sift: To remove large lumps from a dry ingredient such as flour or confectioners' sugar by passing it through a fine mesh. This process also incorporates air into the ingredients, making them lighter.

Simmer: To cook food in a liquid at a low enough temperature that small bubbles begin to break the surface.

Soft-Ball Stage: In candy making, the point at which syrup has cooked long enough that when dropped in cold water it can be shaped into a ball but flattens when removed from the water (234 to 240 degrees F.).

Springform Pan: A pan that is round with high sides. The side rim expands when a clamp is opened, allowing it to separate from the base. This makes it easier to shape and then serve something like a cheesecake.

Steam: To cook over boiling water in a covered pan, this method keeps foods' shape, texture, and nutritional value intact better than methods such as boiling.

Steep: To soak dry ingredients (tea leaves, ground coffee, herbs, spices, etc.) in liquid until the flavor is infused into the liquid.

Thin: To reduce a mixture's thickness with the addition of more liquid.

Unleavened: Baked goods that contain no agents to give them volume, such as baking powder, baking soda, or yeast.

Vinaigrette: A general term referring to any sauce made with vinegar, oil, and seasonings.

Zest: The thin, brightly colored outer part of the rind of citrus fruits. It contains volatile oils, used as a flavoring.

Recipe Index of Huckleberry Delights

312

Reader Feedback Form

Dear Reader,

We are very interested in what our readers think. Please fill in the form below and return it to:

Whispering Pine Press International, Inc.
c/o Huckleberry Delights Cookbook
P.O. Box 214, Spokane Valley, WA 99037-0214
Phone: (509) 928-8700 | Fax: (509) 922-9949
Email: sales@whisperingpinepress.com
Publisher Websites: www.WhisperingPinePress.com
www.WhisperingPinePressBookstore.com
Blog: www.WhisperingPinePressBlog.com

Name: _____

Address: _____

City, St., Zip: _____

Phone/Fax: (____) _____ / (____) _____

Email: _____

Comments/Suggestions: _____

_____

_____

_____

A great deal of care and attention has been exercised in the creation of this book. Designing a great cookbook that is original, fun, and easy to use has been a job that required many hours of diligence, creativity, and research. Although we strive to make this book completely error free, errors and discrepancies may not be completely excluded. If you come across any errors or discrepancies, please make a note of them and send them to our publishing office. We are constantly updating our manuscripts, eliminating errors, and improving quality.

Please contact us at the address above.

About the Cookbook Delights Series

The *Cookbook Delights Series* includes many different topics and themes. If you have a passion for food and wish to know more information about different foods, then this series of cookbooks will be beneficial to you. Each book features a different type of food, such as avocados, strawberries, huckleberries, salmon, vegetarian, lentils, almonds, cherries, coconuts, lemons, and many, many more.

The *Cookbook Delights Series* not only includes cookbooks about individual foods but also includes several holiday-themed cookbooks. Whatever your favorite holiday may be, chances are we have a cookbook with recipes designed with that holiday in mind. Some examples include *Halloween Delights, Thanksgiving Delights, Christmas Delights, Valentine Delights, Mother's Day Delights, St. Patrick's Day Delights,* and *Easter Delights.*

Each cookbook is designed for easy use and is organized into alphabetical sections. Over 250 recipes are included along with other interesting facts, folklore, and history of the featured food or theme. Each book comes with a beautiful full-color cover, ordering information, and a list of other upcoming books in the series.

Note cards, bookmarks, and a daily journal have been printed and are available to go along with each cookbook. You may view the entire line of cookbooks, journals, cards, and posters, by visiting our website at www.whisperingpinepressbookstore or you can email us with your questions and comments to: sales@whisperingpinepress.com.

Please ask your local bookstore to carry these sets of books.

To order, please contact:

Whispering Pine Press International, Inc.

c/o Huckleberry Delights Cookbook
P.O. Box 214, Spokane Valley, WA 99037-0214
Phone: (509) 928-8700 | Fax: (509) 922-9949
Email: sales@whisperingpinepress.com
Publisher Websites: www.WhisperingPinePress.com
www.WhisperingPinePressBookstore.com
Blog: www.WhisperingPinePressBlog.com

We Invite You to Join the Whispering Pine Press International, Inc., Book Club!

Whispering Pine Press International, Inc.

c/o Huckleberry Delights Cookbook
P.O. Box 214, Spokane Valley, WA 99037-0214
Phone: (509) 928-8700 | Fax: (509) 922-9949
Email: sales@whisperingpinepress.com
Publisher Websites: www.WhisperingPinePress.com
www.WhisperingPinePressBookstore.com
Blog: www.WhisperingPinePressBlog.com

Buy 11 books and get the next one free, based on the average price of the first eleven purchased.

How the club works:

Simply use the order form below and order books from our catalog. You can buy just one at a time or all eleven at once. After the first eleven books are purchased, the next one is free. Please add shipping and handling as listed on this form. There are no purchase requirements at any time during your membership. Free book credit is based on the average price of the first eleven books purchased.

Join today! Pick your books and mail in the form today!

Yes! I want to join the Whispering Pine Press International, Inc., Book Club! Enroll me and send the books indicated below.

Title	**Price**
1. _____	_____
2. _____	_____
3. _____	_____
4. _____	_____
5. _____	_____
6. _____	_____
7. _____	_____
8. _____	_____
9. _____	_____
10. _____	_____
11. _____	_____

Free Book Title: _____

Free Book Price: _____ Avg. Price: _____ Total Price: _____

Credit for the free book is based on the average price of the first 11 books purchased.

(Circle one) Check | Visa | MasterCard | Discover | American Express

Credit Card #: _____ Expiration Date: _____

Name: _____

Address: _____

City: _____State: _____Country: _____

Zip/Postal: _____Phone: (_____) _____

Email: _____

Signature_____

316

Whispering Pine Press International, Inc. Fundraising Opportunities

Fundraising cookbooks are proven moneymakers and great keepsake providers for your group. Whispering Pine Press International, Inc., offers a very special personalized cookbook fundraising program that encourages success to organizations all across the USA.

Our prices are competitive and fair. Currently, we offer a special of 100 books with many free features and excellent customer service. Any purchase you make is guaranteed first-rate.

Flexibility is not a problem. If you have special needs, we guarantee our cooperation in meeting each of them. Our goal is to create a cookbook that goes beyond your expectations. We have the confidence and a record that promises continual success.

Another great fundraising program is the *Cookbook Delights Series* Program. With cookbook orders of 50 copies or more, your organization receives a huge discount, making for a prompt and lucrative solution.

We also specialize in assisting group fundraising – Christian, community, nonprofit, and academic among them. If you are struggling for a new idea, something that will enhance your success and broaden your appeal, Whispering Pine Press International, Inc., can help.

For more information, write, phone, or fax to:

Whispering Pine Press International, Inc.
P.O. Box 214
Spokane Valley, WA 99037-0214
Phone: (509) 928-8700 | Fax: (509) 922-9949
Email: sales@whisperingpinepress.com
Publisher Websites: www.WhisperingPinePress.com
www.WhisperingPinePressBookstore.com
Blog: www.WhisperingPinePressBlog.com
Book Website: www.HuckleberryDelights.com
SAN 253-200X

Personalized and/or Translated Order Form for Any Book by Whispering Pine Press International, Inc.

Dear Readers:

If you or your organization wishes to have this book or any other of our books personalized, we will gladly accommodate your needs. For instance, if you would like to change the names of the characters in a book to the names of the children in your family or Sunday school class, we would be happy to work with you on such a project. We can add more information of your choosing and customize this book especially for your family, group, or organization.

We are also offering an option of translating your book into another language. Please fill out the form below telling us exactly how you would like us to personalize your book.

Please send your request to:

Whispering Pine Press International, Inc.
c/o Huckleberry Delights Cookbook
P.O. Box 214, Spokane Valley, WA 99037-0214
Phone: (509) 928-8700 | Fax: (509) 922-9949
Email: sales@whisperingpinepress.com
Publisher Websites: www.WhisperingPinePress.com
www.WhisperingPinePressBookstore.com
Blog: www.WhisperingPinePressBlog.com

Person/Organization placing request: _____

Date_____ Phone: (____) _____

Address_____ Fax: (____) _____

City_____ State_____ Zip: _____

Language of the book: _____

Please explain your request in detail: _____

_____

_____

_____

_____

_____

_____

Huckleberry Delights Cookbook

A Collection of Huckleberry Recipes

How to Order

Get your additional copies of this book by returning an order form and your check, money order, or credit card information to:

Whispering Pine Press International, Inc.
c/o Huckleberry Delights Cookbook
P.O. Box 214, Spokane Valley, WA 99037-0214
Phone: (509) 928-8700 | Fax: (509) 922-9949
Email: sales@whisperingpinepress.com
Publisher Websites: www.WhisperingPinePress.com
www.WhisperingPinePressBookstore.com
Blog: www.WhisperingPinePressBlog.com

Customer Name: _____

Address: _____

City, St., Zip: _____

Phone/Fax: _____

Email: _____

- -

Please send me _____ copies of _____

_____ at $_____ per copy and $4.95 for shipping and handling per book, plus $2.95 each for additional books. Enclosed is my check, money order, or charge my account for $_____.

☐ Check   ☐ Money Order   ☐ Credit Card

(*Circle One*) MasterCard | Discover | Visa | American Express

☐☐☐☐   ☐☐☐☐   ☐☐☐☐   ☐☐☐☐

Expiration Date: _____

_____
Signature

_____
Print Name

Whispering Pine Press International, Inc. Order Form

Gift-wrapping, Autographing, and Inscription
We are proud to offer personal autographing by the author. For a limited time this service is absolutely free!
Gift-wrapping is also available for $4.95 per item.

1. Sold To

Name: _____
Street/Route: _____
_____
City: _____
State: _____ Zip: _____
Country: _____
Gift message: _____
_____
Email address: _____
Daytime Phone: (_ _ _) _ _ _-_ _ _ _
*Necessary for verifying orders
Home Phone: (_ _ _) _ _ _-_ _ _ _
Fax: (_ _ _) _ _ _-_ _ _ _

2. Ship To

[ ] Is this a new or corrected address?
[ ] Alternative Shipping Address
[ ] Mailing Address
Name: _____
Address: _____
_____
City: _____
State: _____ Zip: _____
Country: _____
Email address: _____

3. Items Ordered

ISBN # /Item #	Size	Color	Qty.	Title or Description	Price	Total

4. Method Of Payment

International, Inc. (No Cash or COD's)

[ ] Visa [ ] MasterCard [ ] Discover [ ] American Express [ ] Check/Money Order
Please make it payable to Whispering Pine Press International, Inc. (No Cash or COD's)

Account Number   Expiration Date
_____ / _____
                                  Month Year

[ ][ ][ ] [ ][ ][ ][ ] [ ][ ][ ][ ] [ ][ ][ ][ ]

Signature_____
                    Cardholder's signature
Printed Name_____
                    Please print name of cardholder
Address of Cardholder_____

Subtotal	
Gift wrap $4.95 Each	
For delivery in WA add 8.7% sales tax.	
Shipping See chart at left	
6. Total	

5. Shipping & Handling

Continental US
US Postal Ground: For books please add $4.95 for the first book and $2.95 each for additional books.
All non-book items, add 15% of the Subtotal.
Please allow 1-4 weeks for delivery.
US Postal Air: Please add $15.00 shipping and handling.
Please allow 1-3 days for delivery.
Alaska, Hawaii, and the US Territories By Ship:
Please add 10% shipping and handling
(minimum charge $15.00).

Please
By Air: Please add 12% shipping and handling (minimum charge $15.00).
Please allow 2 –6 weeks for delivery.
International By Ship: Please add 10% shipping and handling (minimum charge $15.00).
Please allow 6-12 weeks for delivery.
By Air: Please add 12% shipping and handling (minimum charge $15.00).
Please allow 2-6 weeks for delivery.
FedEx Shipments: Add $5.00 to the above airmail charges for overnight delivery.

Whispering Pine Press International, Inc.
P.O. Box 214
Spokane Valley, WA 99037-0214 USA
Phone: (509) 928-8700 • Fax: (509) 922-9949
Email: sales@whisperingpinepress.com
Website: www.whisperingpinepress.com

Shop Online:
www.whisperingpinepress.com
Fax orders to: (509) 922-9949

About the Author and Cook

Karen Jean Matsko Hood has always enjoyed cooking, baking, and experimenting with recipes. At this time Hood is working to complete a series of cookbooks that blends her skills and experience in cooking and entertaining. Hood entertains large groups of people and especially enjoys designing creative menus with holiday, international, ethnic, and regional themes.

Hood is publishing a cookbook series entitled the *Cookbook Delights Series*, in which each cookbook emphasizes a different food ingredient or theme. The first cookbook in the series is *Apple Delights Cookbook*. Hood is working to complete another series of cookbooks titled *Hood and Matsko Family Cookbooks*, which includes many recipes handed down from her family heritage and others that have emerged from more current family traditions. She has been invited to speak on talk radio shows on various topics, and favorite recipes from her cookbooks have been prepared on local television programs.

Hood was born and raised in Great Falls, Montana. As an undergraduate, she attended the College of St. Benedict in St. Joseph, Minnesota, and St. John's University in Collegeville, Minnesota. She attended the University of Great Falls in Great Falls, Montana. Hood received a B.S. Degree in Natural Science from the College of St. Benedict and minored in both Psychology and Secondary Education. Upon her graduation, Hood and her husband taught science and math on the island of St. Croix in the U.S. Virgin Islands. Hood has completed postgraduate classes at the University of Iowa in Iowa City, Iowa. In May 2001, she completed her Master's Degree in Pastoral Ministry at Gonzaga University in Spokane, Washington. She has taken postgraduate classes at Lewis and Clark College on the North Idaho college campus in Coeur d'Alene, Idaho, Taylor University in Fort Wayne, Indiana, Spokane Falls Community College, Spokane Community College, Washington State University, University of Washington, and Eastern Washington University. Hood is working on research projects to complete her Ph.D. in Leadership Studies at Gonzaga University in Spokane, Washington.

Hood resides in Greenacres, Washington, along with her husband, many of her sixteen children, and foster children. Her interests include writing, research, and teaching. She previously has volunteered as a court advocate in the Spokane juvenile court system for abused and neglected children. Hood is a literary advocate for youth and adults. Her hobbies include cooking, baking, collecting, photography, indoor and outdoor gardening, farming, and the cultivation of unusual flowering plants and orchids. She enjoys raising several specialty breeds of animals including Babydoll Southdown, Friesen, and Icelandic sheep, Icelandic horses, bichons frisés, cockapoos, Icelandic sheepdogs, a Newfoundland, a Rottweiler, a variety of Nubian and fainting goats, and a few rescue cats. Hood also enjoys bird-watching and finds all aspects of nature precious.

She demonstrates a passionate appreciation of the environment and a respect for all life. She also invites you to visit her websites:

www.KarenJeanMatskoHood.com
www.KarenJeanMatskoHoodBookstore.com
www.KarenJeanMatskoHoodBlog.com
www.KarensKidsBooks.com
www.KarensTeenBooks.com

www.HoodFamilyBlog.com
www.HoodFamily.com

Author's Social Media
Please Follow the Author on **Twitter**: @KarenJeanHood
Friend her on **Facebook**: Karen Jean Matsko Hood Author Fan Page
Google Plus Profile: Karen Jean Matsko Hood
Pinterest.com/KarenJMHood

9 781596 493858